Letts
and
LONSDALE

Revise
AS

19/7/10!

Chemistry

Rob Ritchie

Contents

Chapter 1 Atoms, moles and reactions

Chapter 2 Bonding, structure and the Periodic Table

Chapter 3 Energetics, rates and equilibrium

Chapter 4 Organic chemistry, analysis and the environment

Specification lists

AQA AS Chemistry

MODULE	SPECIFICATION TOPIC	CHAPTER REFERENCE	STUDIED IN CLASS	REVISED	PRACTICE QUESTIONS
AS Unit 1 (M1) Foundation chemistry	Atomic structure	1.1, 1.2, 1.3, 1.4			
	Amount of substance	1.4, 1.5, 1.6			
	Bonding	2.1, 2.2, 2.3, 2.4, 2.5			
	Periodicity	2.6			
	Introduction to organic chemistry	4.1			
	Alkanes	4.2, 4.3			
AS Unit 2 (M2) Chemistry in action	Energetics	3.1, 3.2, 3.3			
	Kinetics	3.4, 3.5			
	Equilibria	3.6, 4.8			
	Redox reactions	1.7			
	Group 7, the Halogens	2.8			
	Group 2, the alkaline earth metals	2.7			
	Extraction of metals	2.9			
	Haloalkanes	4.3, 4.6, 4.8			
	Alkenes	4.4			
	Alcohols	4.5, 4.8			
	Analytical techniques	4.7			

Examination analysis

AS Chemistry comprises two unit tests. All questions are compulsory. Practical and investigative skills will also be assessed.

Unit 1	Structured questions: short and extended answers	1hr 15min test	$33^1/_3\%$
Unit 2	Structured questions: short and extended answers	1hr 45min test	$46^2/_3\%$
Unit 3	Internal assessment of practical and investigative skills		20%

OCR AS Chemistry

MODULE	SPECIFICATION TOPIC	CHAPTER REFERENCE	STUDIED IN CLASS	REVISED	PRACTICE QUESTIONS
AS Unit F321 (M1) *Atoms, bonds and groups*	Atoms	1.1, 1.4			
	Moles and equations	1.5, 1.6			
	Acids	1.8			
	Redox	1.7			
	Electron structure	1.2, 1.3			
	Bonding and structure	2.1, 2.2, 2.3, 2.4, 2.5			
	Periodicity	2.6			
	Group 2	2.7			
	Group 7	2.8			
AS Unit F322 (M2) *Chains, energy and resources*	Basic concepts	4.1, 4.4			
	Alkanes	4.2, 4.3			
	Alkenes	4.4			
	Alcohols	4.5			
	Halogenoalkanes	4.6			
	Modern analytical techniques	4.7			
	Enthalpy changes	3.1, 3.2, 3.3			
	Rates and equilibrium	3.4, 3.5, 3.6			
	Chemistry of the air	3.5, 4.7			
	Green chemistry	4.8			

Examination analysis

AS Chemistry comprises two unit tests. All questions are compulsory.
Practical and investigative skills will also be assessed.

Unit F321	Structured questions: short and extended answers	1hr test	30%
Unit F322	Structured questions: short and extended answers	1hr 45min test	50%
Unit F323	Internal assessment of practical and investigative skills		20%

Salters AS Chemistry

MODULE	SPECIFICATION TOPIC	CHAPTER REFERENCE	STUDIED IN CLASS	REVISED	PRACTICE QUESTIONS
AS Unit F331 (M1) Chemistry for life	Formulae, equations and amount of substance	1.1, 1.4, 1.5, 1.6, 2.1			
	Atomic structure	1.1, 1.2			
	Bonding and structure	2.1, 2.2, 2.5			
	Inorganic chemistry and the Periodic Table	2.6, 2.7, 4.3			
	Modern analytical techniques	1.4, 4.7			
	Energetics	3.1, 3.2, 3.3			
	Kinetics	3.5			
	Organic functional groups	4.1, 4.2			
	Organic reactions	4.3, 4.5			
	Isomerism	4.1			
	Applications	4.2, 4.8			
AS Unit F332 (M2) Chemistry of natural resources	Formulae, equations and amount of substance	1.5, 1.6			
	Atomic structure	1.2			
	Bonding and structure	2.1, 2.2, 2.3, 2.4, 2.5			
	Redox	1.7			
	Inorganic chemistry and the Periodic Table	1.3, 2.1, 2.6, 2.8			
	Organic functional groups	4.1, 4.4, 4.5			
	Organic reactions	4.4, 4.5, 4.8			
	Reaction mechanisms	4.1, 4.3, 4.4, 4.6			
	Isomerism	4.1, 4.4			
	Kinetics	3.4, 3.5			
	Equilibria	3.6			
	Applications of organic chemistry	4.8			
	Modern analytical techniques	4.7, 4.8			

Examination analysis

AS Chemistry comprises two unit tests. All questions are compulsory. Practical and investigative skills will also be assessed.

Unit F331 Structured questions: short and extended answers   1hr 15min test   30%

Unit F332 Structured questions: short and extended answers
One question on an Advance Notice article   1hr 45min test   50%

Unit F333 Internal assessment of practical and investigative skills   20%

Edexcel AS Chemistry

MODULE	SPECIFICATION TOPIC	CHAPTER REFERENCE	STUDIED IN CLASS	REVISED	PRACTICE QUESTIONS
AS Unit 1 (M1) *The core principles of Chemistry*	Formulae, equations and amount of substance	1.1, 1.4, 1.5, 1.6, 4.1			
	Energetics	3.1, 3.2, 3.3			
	Atomic structure and the Periodic Table	1.1, 1.2, 1.3, 1.4, 2.6			
	Bonding	2.1, 2.5			
	Introductory organic chemistry	4.1, 4.2, 4.3, 4.4			
AS Unit 2 (M2) *Application of core principles of Chemistry*	Shapes of molecules and ions	2.2			
	Intermediate bonding and bond polarity	2.3			
	Intermolecular forces	2.4, 2.5			
	Redox	1.7			
	The Periodic Table – Groups 2 and 7	2.7, 2.8			
	Kinetics	3.4, 3.5			
	Chemical equilibria	3.6			
	Organic chemistry	4.5, 4.6			
	Mechanisms	4.1, 4.3, 4.4, 4.6, 4.8			
	Mass spectra and IR	4.7, 4.8			
	Green chemistry	4.8			

Examination analysis

AS Chemistry comprises two unit tests. All questions are compulsory.
Practical and investigative skills will also be assessed.

Unit 1	Objective questions; Structured questions: short and extended answers	1hr 15min test	40%
Unit 2	Objective questions; Structured questions: short and extended answers contemporary context questions	1hr 15min test	40%
Unit 3	Internal assessment of practical and investigative skills		20%

WJEC AS Chemistry

MODULE	SPECIFICATION TOPIC	CHAPTER REFERENCE	STUDIED IN CLASS	REVISED	PRACTICE QUESTIONS
AS Unit CH1 **(M1)** Controlling and using chemical change	Basic ideas about atoms	1.1, 1.2, 1.3			
	Chemical calculations	1.4, 1.5, 1.6, 4.1			
	Chemical equilibrium and acid-base reactions	1.6, 1.8, 3.6			
	Energetics	3.1, 3.2, 3.3			
	Kinetics	3.4, 3.5			
	Application of principles	4.8			
AS Unit CH2 **(M2)** Properties, structure and bonding	Chemical bonding	2.1, 2.3			
	Forces between molecules	2.4			
	Shapes of molecules	2.2			
	Solubility of compounds in water	2.5			
	Solid structures	2.5			
	The Periodic Table	1.7, 2.6			
	Trends in properties of the elements of the s-block and Group 7	2.7, 2.8			
	Organic compounds and their reactions	1.6, 4.1, 4.2, 4.4, 4.5, 4.6			
	Hydrocarbons	4.1, 4.2, 4.3, 4.4			
	Halogenoalkanes	4.3, 4.6			
	Alcohols	4.5			
	Analytical techniques	4.7			

Examination analysis

AS Chemistry comprises two unit tests. All questions are compulsory.
Practical and investigative skills will also be assessed.

Unit 1	Objective questions; Structured questions: short and extended answers	1hr 30min test	40%
Unit 2	Objective questions; Structured questions: short and extended answers	1hr 30min test	40%
Unit 3	Internal assessment of practical and investigative skills		20%

$$CH_4(g) + 3O_2(g)$$
$$\rightarrow CO_2 + H_2O$$

CCEA AS Chemistry

MODULE	SPECIFICATION TOPIC	CHAPTER REFERENCE	STUDIED IN CLASS	REVISED	PRACTICE QUESTIONS
AS Unit 1 (M1) Basic concepts in physical and inorganic chemistry	Formulae, equations and amounts of substance	1.5, 1.6, 1.8			
	Atomic structure	1.1, 1.2, 1.3, 1.4, 2.10			
	Bonding and structure	2.1, 2.3, 2.5			
	Shapes of molecules and ions	2.2			
	Intermolecular forces	2.4, 2.5			
	Redox	1.7			
	The Periodic Table	2.6			
	Group VII	2.8			
	Titrations	1.5			
AS Unit 2 (M2) Further physical and inorganic chemistry and introduction to organic chemistry	Formulae and formulae of a substance	1.5, 1.6, 4.1			
	Nomenclature and isomerism in organic compounds	4.1, 4.2			
	Hydrocarbons – alkanes	4.1, 4.2, 4.3, 3.5			
	Hydrocarbons – alkenes	4.4			
	Halogenoalkanes	4.6			
	Alcohols	4.5			
	Infra-red spectroscopy	4.7			
	Energetics	3.1, 3.2, 3.3			
	Equilibrium	3.6			
	Kinetics	3.4, 3.5			
	Group II elements and their compounds	2.7			
	Qualitative analysis	2.10			

Examination analysis

AS Chemistry comprises two unit tests. All questions are compulsory.
Practical and investigative skills will also be assessed.

Unit 1	Objective questions; Structured questions: short and extended answers	1hr 30min test	35%
Unit 2	Objective questions; Structured questions: short and extended answers contemporary context questions	1hr 30min test	35%
Unit 3	Internal assessment of practical and investigative skills		30%

AS/A2 Level Chemistry courses

AS and A2

All Chemistry A Level courses being studied from September 2000 are in two parts, with three separate modules in each part. Students first study the AS (Advanced Subsidiary) course. Some will then go on to study the second part of the A Level course, called A2. Advanced Subsidiary is assessed at the standard expected halfway through an A Level course: i.e., between GCSE and Advanced GCE. This means that new AS and A2 courses are designed so that difficulty steadily increases:

* AS Chemistry builds from GCSE science
* A2 Chemistry builds from AS Chemistry.

How will you be tested?

Assessment units

For AS Chemistry, you will be tested by three assessment units. For the full A Level in Chemistry, you will take a further three units. AS Chemistry forms 50% of the assessment weighting for the full A Level.

One of the units in AS and in A2 is practically based. You will take two theory units in each of AS and A2. Each unit can normally be taken in either January or June. Alternatively, you can study the whole course before taking any of the unit tests. There is a lot of flexibility about when exams can be taken and the diagram below shows just some of the ways that the assessment units may be taken for AS and A Level Chemistry.

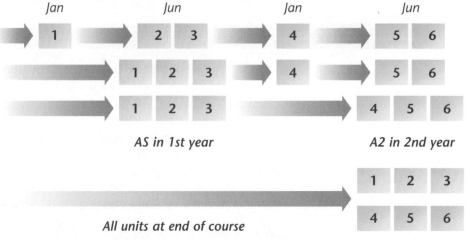

If you are disappointed with a module result, you can resit each module once. You will need to be very careful about when you take up a resit opportunity because you will have only one chance to improve your mark. The higher mark counts.

A2 and Synoptic assessment

After having studied AS Chemistry, you may wish to continue studying Chemistry to A Level. For this you will need to take three further units of Chemistry at A2. Similar assessment arrangements apply except some units, those that draw together different parts of the course in a synoptic assessment, have to be assessed at the end of the course.

Coursework

Coursework may form part of your A Level Chemistry course, depending on which specification you study. Where students have to undertake coursework, it is usually for the assessment of practical skills but this is not always the case. See pages 4–9.

Key Skills

To gain the key skills qualification, which is equivalent to an AS Level, you will need to collect evidence together in a 'portfolio' to show that you have attained a satisfactory level in Communication, Application of number and Information technology. You will also need to take a formal testing in each key skill. You will have many opportunities during AS Chemistry to develop your key skills.

What skills will I need?

The assessment objectives for AS Chemistry are shown below.

Knowledge with understanding

- recall of facts, terminology and relationships
- understanding of principles and concepts
- drawing on existing knowledge to show understanding of the responsible use of chemistry in society
- selecting, organising and presenting information clearly and logically

Application of knowledge and understanding, analysis and evaluation

- explaining and interpreting principles and concepts
- interpreting and translating, from one form into another, data presented as continuous prose or in tables, diagrams and graphs
- carrying out relevant calculations
- applying knowledge and understanding to familiar and unfamiliar situations
- assessing the validity of chemical information, experiments, inferences and statements

Experimental and investigative skills

Chemistry is a practical subject and part of the assessment of AS Chemistry will test your practical skills. You will be assessed on three main skills:

- implementing
- analysing evidence and drawing conclusions
- evaluating evidence and procedures.

The skills may be assessed in the context of separate practical exercises, although more than one skill could be assessed in any one exercise. They may also be assessed all together in the context of a single 'whole investigation'. An investigation may be set by your teacher or you may be able to pursue an investigation of your choice. Alternatively, you may take a practical examination.

You will receive guidance about how your practical skills will be assessed from your teacher. This study guide concentrates on preparing you for the written examinations testing the subject content of AS Chemistry.

Different types of questions in AS examinations

In AS Chemistry examinations, different types of question are used to assess your abilities and skills. Unit tests mainly use structured questions requiring both short answers and more extended answers.

Short-answer questions

A short-answer question may test recall or it may test understanding. Short-answer questions normally have space for the answers printed on the question paper.

Here are some examples (the answers are shown in *italics*):

What is meant by isotopes?

Atoms of the same element with different masses.

Calculate the amount (in mol) of H_2O in 4.5 g of H_2O.

1 mol H_2O has a mass of 18 g. ∴ 4.5 g of H_2O contains 4.5/18 = 0.25 mol H_2O.

Structured questions

Structured questions are in several parts. The parts usually have a common context and they often become progressively more difficult and more demanding as you work your way through the question. A structured question may start with simple recall, then test understanding of a familiar or an unfamiliar situation.

Most of the practice questions in this book are structured questions, as this is the main type of question used in the assessment of AS Chemistry.

When answering structured questions, do not feel that you have to complete one question before starting the next. The further you are into a question, the more difficult the marks are to obtain. If you run out of ideas, go on to the next question. You need to respond to as many parts of questions on an exam paper as possible. You will not score well if you spend so long trying to perfect the first questions that you do not have time to attempt later questions.

Here is an example of a structured question that becomes progressively more demanding.

(a) Write down the atomic structure of the two isotopes of potassium: ^{39}K and ^{41}K.

   (i) ^{39}K ...19... protons; ...20... neutrons; ...19... electrons. ✔

   (ii) ^{41}K ...19... protons; ...22... neutrons; ...19... electrons. ✔   [2]

(b) A sample of potassium has the following percentage composition by mass: ^{39}K: 92%; ^{41}K: 8% Calculate the relative atomic mass of the potassium sample.

   $92 \times 39/100 + 8 \times 41/100 = 39.16$ ✔   [1]

(c) What is the electronic configuration of a potassium atom?

   $1s^2 2s^2 2p^6 3s^2 3p^6 4s^1$ ✔   [1]

(d) The second ionisation energy of potassium is much larger than its first ionisation energy.

(i) Explain what is meant by the *first ionisation energy* of potassium.

> The energy required to remove an electron ✓ from each atom in 1 mole ✓ of gaseous atoms ✓

(ii) Why is there a large difference between the values for the first and the second ionisation energies of potassium?

> The 2nd electron removed is from a different shell ✓ which is closer to the nucleus and experiences more attraction from the nucleus. ✓ This outermost electron experiences less shielding from the nucleus because there are fewer inner electron shells than for the 1st ionisation energy. ✓   [6]

Extended answers

In AS Chemistry, questions requiring more extended answers may form part of structured questions or may form separate questions. They may appear anywhere on the paper and will typically have between 5 and 10 marks allocated to the answers as well as several lines of answer space. These questions are also often used to assess your abilities to communicate ideas and put together a logical argument.

The correct answers to extended questions are often less well-defined than to those requiring short answers. Examiners may have a list of points for which credit is awarded up to the maximum for the question.

An example of a question requiring an extended answer is shown below.

Magnesium oxide and sulfur trioxide are both solids at $10\,°C$. Magnesium oxide has a melting point of $2852\,°C$. Sulfur trioxide has a melting point of $17\,°C$. Explain, in terms of structure and bonding, why these two compounds have such different melting points.   [7]

Points that the examiners might look for include:

> Magnesium oxide has a giant lattice structure ✓ containing ionic bonds. ✓ These strong forces need to be broken during melting to allow ions to move free of the rigid lattice. ✓ This requires a large amount of energy supplied by a high temperature. ✓
>
> Sulfur trioxide has a simple molecular structure ✓ held together by van der Waals' forces. ✓ These weak forces need to be broken during melting to allow molecules to move free of the rigid lattice. ✓ This requires a small amount of energy supplied by a low temperature. ✓

8 marking points → [7]

In this type of response, there may be an additional mark for a clear, well-organised answer, using specialist terms. In addition, marks may be allocated for legible text with accurate spelling, punctuation and grammar.

Other types of questions

Free-response and open-ended questions allow you to choose the context and to develop your own ideas. These are little used for assessing AS Chemistry but, if you decide to take the full A Level in Chemistry, you will encounter this type of question during synoptic assessment.

Multiple-choice or objective questions require you select the correct response to the question from a number of given alternatives. These are rarely used for assessing AS Chemistry, although it is possible that some short-answer questions will use a multiple-choice format.

Exam technique

Links from GCSE

Advanced Subsidiary Chemistry builds from grade CC in GCSE Science and GCSE Additional Science (combined), or GCSE Chemistry. This study guide has been written so that you will be able to tackle AS Chemistry from a GCSE Science background.

You should not need to search for important Chemistry from GCSE Science because this has been included where needed in each chapter. If you have not studied Science for some time, you should still be able to learn AS Chemistry using this text alone.

What are examiners looking for?

Examiners use instructions to help you to decide the length and depth of your answer.

If a question does not seem to make sense, you may have misread it – read it again!

State, define or list

This requires a short, concise answer, often recall of material that can be learnt by rote.

Explain, describe or discuss

Some reasoning or some reference to theory is required, depending on the context.

Outline

This implies a short response, almost a list of sentences or bullet points.

Predict or deduce

You are not expected to answer by recall but by making a connection between pieces of information.

Suggest

You are expected to apply your general knowledge to a 'novel' situation, one which you have not directly studied during the AS Chemistry course.

Calculate

This is used when a numerical answer is required. You should always use units in quantities and significant figures should be used with care.

Look to see how many significant figures have been used for quantities and give your answer to this degree of accuracy.

If the question uses three significant figures, then give your answer in three significant figures also.

Some dos and don'ts

Dos

Do answer the question

No credit can be given for good Chemistry that is irrelevant to the question.

Do use the mark allocation to guide how much you write

Two marks are awarded for two valid points – writing more will rarely gain more credit and could mean wasted time or even contradicting earlier valid points.

Do use diagrams, equations and tables in your responses

Even in 'essay-type' questions, these offer an excellent way of communicating chemistry.

Do write legibly

An examiner cannot give marks if the answer cannot be read.

Do write using correct spelling and grammar. Structure longer essays carefully

Marks are now awarded for the quality of your language in exams.

Don'ts

Don't fill up any blank space on a paper

In structured questions, the number of dotted lines should guide the length of your answer.

If you write too much, you waste time and may not finish the exam paper. You also risk contradicting yourself.

Don't write out the question again

This wastes time. The marks are for the answer!

Don't contradict yourself

The examiner cannot be expected to choose which answer is intended. You could lose a hard-earned mark, e.g.

A covalent bond is a shared pair of electrons ✓ bonded by the electrostatic attraction between ions. ✗

Don't spend too much time on a part that you find difficult

You may not have enough time to complete the exam. You can always return to a difficult part if you have time at the end of the exam.

What grade do you want?

Everyone would like to improve their grades but you will only manage this with a lot of hard work and determination. You should have a fair idea of your natural ability and likely grade in Chemistry and the hints below offer advice on improving that grade.

For a Grade A

You will need to be a very good all-rounder.

- You must go into every exam knowing the work extremely well.
- You must be able to apply your knowledge to new, unfamiliar situations.
- You need to have practised many, many exam questions so that you are ready for the type of question that will appear.

The exams test all areas of the syllabus and any weaknesses in your Chemistry will be found out. There must be no holes in your knowledge and understanding. For a Grade A, you must be competent in all areas.

At A level, you also have the opportunity to achieve an A* grade.

For a Grade C

You must have a reasonable grasp of Chemistry but you may have weaknesses in several areas and you may be unsure of some of the reasons for the Chemistry.

- Many Grade C candidates are just as good at answering questions as the Grade A students but holes and weaknesses often show up in just some topics.
- To improve, you will need to master your weaknesses and you must prepare thoroughly for the exam. You must become a better all-rounder.

For a Grade E

You cannot afford to miss the easy marks. Even if you find Chemistry difficult to understand and would be happy with a Grade E, there are plenty of questions in which you can gain marks.

- You must memorise all definitions.
- You must practise exam questions to give yourself confidence that you do know some Chemistry. In exams, answer the parts of questions that you know first. You must not waste time on the difficult parts. You can always go back to these later.
- The areas of Chemistry that you find most difficult are going to be hard to score on in exams. Even in the difficult questions, there are still marks to be gained. Show your working in calculations because credit is given for a sound method. You can always gain some marks if you get part of the way towards the solution.

What marks do you need?

The table below shows how your average mark is transferred into a grade.

average	80%	70%	60%	50%	40%
grade	A	B	C	D	E

Four steps to successful revision

Step 1: Understand

- Study the topic to be learned slowly. Make sure you understand the logic or important concepts.
- Mark up the text if necessary – underline, highlight and make notes.
- Re-read each paragraph slowly.

GO TO STEP 2

Step 2: Summarise

- Now make your own revision note summary:
  What is the main idea, theme or concept to be learned?
  What are the main points? How does the logic develop?
  Ask questions: Why? How? What next?
- Use bullet points, mind maps, patterned notes.
- Link ideas with mnemonics, mind maps, crazy stories.
- Note the title and date of the revision notes
  (e.g. Chemistry: Atomic structure, 3rd March).
- Organise your notes carefully and keep them in a file.

This is now in **short-term memory**. You will forget 80% of it if you do not go to Step 3.
GO TO STEP 3, but first take a 10 minute break.

Step 3: Memorise

- Take 25 minute learning 'bites' with 5 minute breaks.
- After each 5 minute break test yourself:
  Cover the original revision note summary
  Write down the main points
  Speak out loud (record on tape)
  Tell someone else
  Repeat many times.

The material is well on its way to **long-term memory**.
You will forget 40% if you do not do step 4. **GO TO STEP 4**

Step 4: Track/Review

- Create a Revision Diary (one A4 page per day).
- Make a revision plan for the topic, e.g. 1 day later, 1 week later, 1 month later.
- Record your revision in your Revision Diary, e.g.
  Chemistry: Atomic Structure, 3rd March 25 minutes
  Chemistry: Atomic Structure, 5th March 15 minutes
  Chemistry: Atomic Structure, 3rd April 15 minutes
  ... and then at monthly intervals.

Chapter 1

Atoms, moles and reactions

The following topics are covered in this chapter:

- Atoms and isotopes
- The electron structure of the atom
- Experimental evidence for electron structures
- Atomic mass

- The mole
- Formulae, equations and reacting quantities
- Redox reactions
- Acids, bases and salts

1.1 Atoms and isotopes

After studying this section you should be able to:

- recall the relative charge and mass of a proton, neutron and electron
- explain the existence of isotopes
- use atomic number and mass number to determine atomic structure
- describe characteristics of radioactivity

LEARNING SUMMARY

What is an atom?

AQA	M1	OCR	M1
EDEXCEL	M1	SALTERS	M1
CCEA	M1	WJEC	M1

An atom is the smallest part of an element that can exist on its own. Atoms are so tiny that there are more atoms in a full stop than there are people in the world. It is now possible to see individual atoms by using the most modern and powerful microscopes. However, nobody has yet seen *inside* an atom and we must devise models for the structure of an atom from experimental evidence.

electron shells

nucleus (protons and neutrons)

Sub-atomic particles

Although there are various models for atomic structure, chemists use a model in which an atom is composed of three **sub-atomic particles**: protons, neutrons and electrons.

In this model, protons and neutrons form the nucleus at the centre of the atom with electrons orbiting in shells. Compared with the total volume of an atom, the nucleus is tiny and extremely dense. Most of an atom is empty, made up of the space between the nucleus and the electron shells.

All matter is made up from the chemical elements and 116 are now known (although others will inevitably be discovered). The number of protons in the nucleus distinguishes the atoms of each element.

> An important principle of chemistry is the link between an element and the number of protons in its atoms.

> **KEY POINT**
>
> Each atom of an element has the same number of protons.
> The number of protons in an atom of an element is called the **atomic number**, Z.

Properties of protons, neutrons and electrons

> All atoms of hydrogen contain 1 proton.
> All atoms of carbon contain 6 protons.
> All atoms of oxygen contain 8 protons.

Some properties of protons, neutrons and electrons are shown in the table below.

particle	relative mass	relative charge
proton, p	1	1+
neutron, n	1	0
electron, e	1/1840	1–

A proton has virtually the same mass as a neutron and, for most of chemistry, their masses can be assumed to be identical.

An electron has negligible mass compared with the mass of a proton or a neutron. In most of chemistry, the mass of an electron can be ignored.

The charge on a proton is opposite to that of an electron but each charge has the same magnitude. An atom is electrically neutral. To balance out the charges, an atom must have the same number of protons as electrons.

All atoms of hydrogen contain:
1 proton and 1 electron.

All atoms of carbon contain:
6 protons and 6 electrons.

All atoms of oxygen contain:
8 protons and 8 electrons.

> **KEY POINT**
> An atom is electrically neutral.
> An atom contains the same number of protons as electrons.

Isotopes

Without neutrons, the nucleus would just contain positively-charged protons. Like-charges repel, and a nucleus containing just protons would fly apart! Strong nuclear forces act between protons and neutrons and these hold the nucleus together. Neutrons can be thought of as 'nuclear glue'.

In each atom of an element, the number of protons in the nucleus is fixed. However, most elements contain atoms with different numbers of neutrons. These atoms are called **isotopes** and, because they have different numbers of neutrons, they also have different masses.

> **KEY POINT**
> Isotopes are atoms of the **same** element with **different** masses.
> Isotopes of an element have the **same** number of **protons and electrons**.
> Each isotope has a **different** number of **neutrons** in the nucleus.
> The combined number of protons and neutrons in an isotope of an element is called the **mass number**, A.

The isotopes of an element react in the same way. This is because chemical reactions involve electrons – neutrons make no difference.

Chemists represent the nucleus of an isotope in a special way.

mass number
(protons + neutrons)

$^A_Z X$

atomic number
(protons)

The isotopes of carbon

Carbon exists as three isotopes, $^{12}_6C$, $^{13}_6C$, $^{14}_6C$. In all but the most accurate work, it is reasonable to assume that the relative mass of an isotope is equal to its mass number. It is easy to work out the atomic structure of an isotope from the representation above.

isotope	atomic number	mass number	protons	neutrons	electrons
$^{12}_6C$	6	12	6	6	6
$^{13}_6C$	6	13	6	7	6
$^{14}_6C$	6	14	6	8	6

Because all carbon isotopes have 6 protons, it is common practice to omit the atomic number, and $^{12}_6C$ is often shown as ^{12}C, or even as carbon-12.

Progress check

How many protons, neutrons and electrons are in the following isotopes?
1 ^7_3Li; 2 $^{23}_{11}Na$; 3 $^{19}_9F$; 4 $^{27}_{13}Al$; 5 $^{55}_{26}Fe$.

5 $^{55}_{26}Fe$: 26p, 29n, 26e.
4 $^{27}_{13}Al$: 13p, 14n, 13e;
3 $^{19}_9F$: 9p, 10n, 9e;
2 $^{23}_{11}Na$: 11p, 12n, 11e
1 ^7_3Li: 3p, 4n, 3e

Radioactivity

SALTERS · M1

WJEC · M1

The nuclei of some atoms are unstable and these atoms are **radioactive**.

Nuclear fission

During nuclear **fission**, the nuclei of radioactive atoms decay, emitting α and β-particles and γ-radiation.

An α-particle is a helium nucleus.

- An α-**particle**, ^4_2He or $^4_2\alpha$, consists of 2p and 2n.
- A β-**particle**, $^0_{-1}\text{e}$ or $^0_{-1}\beta$, is an electron emitted from the nucleus.
- γ-**radiation** is high energy, electromagnetic radiation that often accompanies the emission of α and β-particles.

Notice that α and β are particles; γ is electromagnetic radiation.

α-particles are the least penetrating form of radioactivity and can be stopped by paper.

γ-radiation is the most penetrating radioactivity.

	α	β	γ
relative charge	+2	–1	0
relative mass	4	0.00055	0
range in air	a few cm	a few m	long
deflection by electric field	small	large	none

During α-**decay**,

- the mass number goes **down** by 4
- the atomic number goes **down** by 2.

e.g.
$$^{238}_{92}\text{U} \longrightarrow \, ^{234}_{90}\text{Th} + ^4_2\text{He}$$

During β-**decay**,

- a neutron changes into a proton: $^1_0\text{n} \longrightarrow \, ^1_1\text{p} + ^0_{-1}\text{e}$
- the mass number stays the **same**
- the atomic number goes **up** by 1.

On either side of a nuclear equation, the atomic numbers and mass numbers must balance.

e.g.
$$^{14}_6\text{C} \longrightarrow \, ^{14}_7\text{N} + ^0_{-1}\text{e}$$

Half-life

Half-life is the time taken for half the radioactive nuclei in a sample to decay. A radioactive isotope has a fixed half-life.

The half-life of ^{90}Sr is 28 years.

Use of radioisotopes.

Radioisotopes can be used as 'tracers' in the body and dating of archaeological and geological material.

The half-life of 'tracers' must be of an appropriate length to allow detection but not cause undue damage.

- In one half-life (28 years), 4 g of ^{90}Sr decays to 2 g   ($\frac{1}{2}$ original)
- In two half-lives (56 years), 1 g of ^{90}Sr remains   ($\frac{1}{4}$ original)
- In three half-lives (84 years), 0.5 g of ^{90}Sr remains   ($\frac{1}{8}$ original)

Fusion

During nuclear **fusion**, nuclei of light atoms join to give heavier nuclei.

Nuclear fusion of hydrogen isotopes takes place in the nuclei of stars such as the Sun:

e.g.
$$^1_1\text{H} + ^2_1\text{H} \longrightarrow \, ^3_2\text{He} + \gamma$$
$$^3_1\text{H} + ^2_1\text{H} \longrightarrow \, ^4_2\text{He} + ^1_0\text{n}$$

1.2 The electron structure of the atom

After studying this section you should be able to:

- understand the relationship between shells, sub-shells and orbitals
- understand how atomic orbitals are filled
- describe electron structure in terms of shells and sub-shells

LEARNING SUMMARY

Energy levels or 'shells'

AQA	M1	OCR	M1
EDEXCEL	M1	SALTERS	M1
CCEA	M1	WJEC	M1

energy

$n = 4$ $32e^-$ ___

$n = 3$ $18e^-$ ___

$n = 2$ $8e^-$ ___

$n = 1$ $2e^-$ ___

Electron energy levels are like a ladder and are filled from the bottom up.

Notice that the gap between successive energy levels becomes less with increasing energy.

You can imagine a model of an atom with electrons orbiting in shells around the nucleus. The electrons in each successive shell have an orbit further away from the nucleus. The further a shell is from the nucleus, the greater the shell's energy level. Each energy level is given a number called the principal quantum number, n. The shell closest to the nucleus has an energy level with $n = 1$, then $n = 2$ and so on.

The number of electrons that can occupy the first four energy levels is shown below:

n	shell	electrons
1	1st shell	2
2	2nd shell	8
3	3rd shell	18
4	4th shell	32

The principal quantum shell, n, is the shell number.

If you look closely, you can see a pattern. The number of electrons that can occupy a shell is $2n^2$. ·

Using this model, electrons can be placed into available shells, starting with the lowest energy level. Each shell must be full before the next starts to fill. The table below shows how the shells are filled for the first 11 elements in the Periodic Table.

element	atomic number	electrons		
		$n = 1$	$n = 2$	$n = 3$
H	1	1		
He	2	2		
Li	3	2	1	
Be	4	2	2	
B	5	2	3	
C	6	2	4	
N	7	2	5	
O	8	2	6	
F	9	2	7	
Ne	10	2	8	
Na	11	2	8	1

This model breaks down as the $n = 3$ energy level is filled because each shell consists of sub-shells.

Sub-shells and orbitals

AQA	M1	OCR	M1
EDEXCEL	M1	SALTERS	M2
CCEA	M1	WJEC	M1

A more advanced model of electron structure is used in which each shell is made up of **sub-shells**.

Sub-shells

There are different types of sub-shell: s, p, d and f. Each type of sub-shell can hold a different number of electrons:

sub-shell	electrons
s	2
p	6
d	10
f	14

The table below shows the shells and sub-shells for the first four principal quantum numbers.

> Notice the labelling. The s sub-shell in the 2nd shell is labelled 2s.

n	shell	sub-shell				total number of electrons	
1	1st shell	1s				2	= 2
2	2nd shell	2s	2p			2 + 6	= 8
3	3rd shell	3s	3p	3d		2 + 6 + 10	= 18
4	4th shell	4s	4p	4d	4f	2 + 6 + 10 + 14	= 32

- Each successive shell contains a new type of sub-shell.
- The 1st shell contains 1 sub-shell, the second shell contains 2 sub-shells, and so on.

Orbitals

How do the electrons fit into the sub-shells? Mathematicians have worked out that electrons occupy negative charge clouds called **orbitals** and these make up each sub-shell.

> **KEY POINT**
> - An orbital can hold up to two electrons.
> - Each type of sub-shell has different orbitals: s, p, d and f.

The table below shows how electrons fill the orbitals in each sub-shell.

sub-shell	orbitals	electrons
s	1	1 x 2 = 2
p	3	3 x 2 = 6
d	5	5 x 2 = 10
f	7	7 x 2 = 14

s-orbitals

> One s-orbital

- An s-orbital has a spherical shape.

p-orbitals

> Three p-orbitals

- A p-orbital has a 3-dimensional dumb-bell shape.
- There are three p-orbitals, p_x, p_y and p_z, at right angles to one another.

> Orbitals are regions around a nucleus that have electron density.

Three p-orbitals

giving

d-orbitals and f-orbitals

The structures of d and f-orbitals are more complex.

Five d-orbitals
Seven f-orbitals

- There are five d-orbitals.
- There are seven f-orbitals.

How do two electrons fit into an orbital?

'Electrons in a box'

allowed

$\uparrow\downarrow$ $\quad$ $\downarrow\uparrow$

not allowed

$\uparrow\uparrow$ $\quad$ $\downarrow\downarrow$

Electrons are negatively charged and so they repel one another. An electron also has a property called **spin**. The two electrons in an orbital have opposite spins, helping to counteract the natural repulsion between their negative charges.

> - Chemists often represent an orbital as a box that can hold up to 2 electrons.
> - Each electron is shown as an arrow, indicating its spin: either $\uparrow$ or $\downarrow$.
> - Within an orbital, the electrons must have **opposite spins**.
>
> **KEY POINT**

Filling the sub-shells

AQA	M1	OCR	M1
EDEXCEL	M1	SALTERS	M2
CCEA	M1	WJEC	M1

Within a shell, the sub-shell energies are in the order: s, p, d and f.

Note that the 4s sub-shell is at a lower energy than the 3d sub-shell.

The 4s sub-shell fills before the 3d sub-shell.

Sub-shells have different energy levels. The diagram below shows the relative energies for the sub-shells in the first four shells.

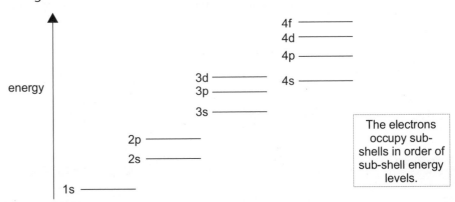

> The electrons occupy sub-shells in order of sub-shell energy levels.

> - Shells and sub-shells are occupied in energy-level order.
> - Electrons occupy orbitals singly before pairing begins to prevent any repulsion caused by pairing.
>
> **KEY POINT**

The 2s orbital is occupied before the 2p orbitals because it is at a lower energy.

Note that the 2p orbitals are occupied singly before pairing begins.

The diagram below shows how electrons occupy orbitals from boron to oxygen.

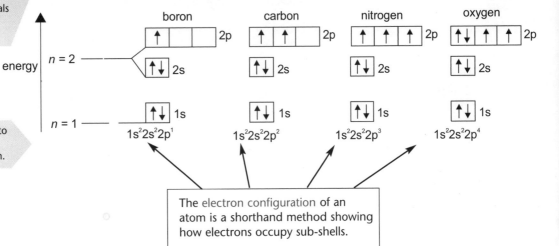

Use these 4 examples to see how electron configuration is written.

The **electron configuration** of an atom is a shorthand method showing how electrons occupy sub-shells.

The diagram below shows how the sub-shells are filled in an atom of potassium. Notice why the 4s sub-shell starts to fill before the 3d sub-shell.

The 4s sub-shell has a lower energy than the 3d sub-shell.
The 4s sub-shell is occupied **before** the 3d sub-shell.

Sub-shells and the Periodic Table

AQA	M1	OCR	M1
EDEXCEL	M1	SALTERS	M2
CCEA	M1	WJEC	M1

The Periodic Table is structured in blocks of 2, 6, 10 and 14, linked to sub-shells. The order of sub-shell filling can be seen by dividing the Periodic Table into blocks.

You should be able to work out the electron configuration for any element in the first four periods, i.e. up to krypton ($Z = 36$).

				1s			1s
s-block							**p-block**
2s							2p
3s			**d-block**				3p
4s			3d				4p
5s			4d				5p
6s			5d				6p
7s							

f-block
4f
5f

Simplifying electron configurations

The similar electron configurations within a group of the Periodic Table can be emphasised with a simpler representation in terms of the previous noble gas.

The electron configurations for the elements in Group 1 are shown below:

Li: $1s^2 2s^1$ or   [He]$2s^1$

Na: $1s^2 2s^2 2p^6 3s^1$ or   [Ne]$3s^1$

K: $1s^2 2s^2 2p^6 3s^2 3p^6 4s^1$ or   [Ar]$4s^1$

Progress check

1 Write the full electron configuration in terms of sub-shells for:
(a) C; (b) Al; (c) Ca; (d) Fe; (e) Br.

1 (a) C: $1s^2 2s^2 2p^2$
(b) Al: $1s^2 2s^2 2p^6 3s^2 3p^1$
(c) Ca: $1s^2 2s^2 2p^6 3s^2 3p^6 4s^2$
(d) Fe: $1s^2 2s^2 2p^6 3s^2 3p^6 3d^6 4s^2$
(e) Br: $1s^2 2s^2 2p^6 3s^2 3p^6 3d^{10} 4s^2 4p^5$

1.3 Experimental evidence for electron structures

After studying this section you should be able to:

- *recall the definitions for first and successive ionisation energies*
- *understand the factors affecting the sizes of ionisation energies*
- *show how ionisation energies provide evidence for shells and sub-shells*
- *understand evidence for electron structure from electron emission and absorption spectra*

LEARNING SUMMARY

Ions

AQA	M1	OCR	M1
EDEXCEL	M1	SALTERS	M1
CCEA	M1	WJEC	M1

Positive ions form when electrons are lost.

Negative ions form when electrons are gained.

An atom can either lose or gain electrons to form an **ion**: a charged atom.

A positive ion is formed when an atom loses electrons. For example, a lithium atom forms a **positive ion** with a 1+ charge by **losing** an electron:

$$Li \longrightarrow Li^+ + e^-$$
$$(3p^+, 4n, \mathbf{3e^-}) \quad\quad (3p^+, 4n, \mathbf{2e^-})$$

A negative ion is formed when an atom gains electrons. For example, an oxygen atom forms a **negative ion** with a 2– charge by **gaining** two electrons:

$$O + 2e^- \longrightarrow O^{2-}$$
$$(6p^+, 6n, \mathbf{6e^-}) \quad\quad\quad\quad (6p^+, 6n, \mathbf{8e^-})$$

Ionisation energy

AQA	M1	OCR	M1
EDEXCEL	M1	SALTERS	M2
CCEA	M1	WJEC	M1

Ionisation energy measures the ease with which electrons are lost in the formation of positive ions. An element has as many ionisation energies as there are electrons.

> **KEY POINT**
>
> The **first** *ionisation energy* of an element is the energy required to remove **1 electron** from each atom in **1 mole** of gaseous atoms to form 1 mole of gaseous 1+ ions.

The equation representing the first ionisation energy of sodium is shown below.

$$Na(g) \longrightarrow Na^+(g) + e^- \quad\quad \text{1st ionisation energy} = +496 \text{ kJ mol}^{-1}$$

Factors affecting ionisation energy

Electrons are held in their shells by attraction from the nucleus. The first electron lost will be from the highest occupied energy level. This electron experiences least attraction from the nucleus.

Three factors affecting the size of this attraction are shown below.

These factors are very important and help to explain many chemical ideas throughout the course.

Learn them!

> **Atomic radius**
> The greater the distance between the nucleus and the outer electrons, the less the attractive force. Attraction falls rapidly with increasing distance and so this factor is very important and has a big effect.
>
> **Nuclear charge**
> The greater the number of protons in the nucleus, the greater the attractive force.
>
> **Electron shielding or 'screening'**
> The outer shell electrons are repelled by any inner shells between the electrons and the nucleus.
>
> This repelling effect, called electron shielding or screening, reduces the overall attractive force experienced by the outer electrons.
>
> **KEY POINT**

Evidence for shells

EDEXCEL	M1	SALTERS	M2
CCEA	M1	WJEC	M1
OCR	M1		

> **KEY POINT**
>
> The **second ionisation energy** of an element is the energy required to remove **1 electron** from each atom in **1 mole** of gaseous 1+ ions to form 1 mole of gaseous 2+ ions.

This pattern shows that a sodium atom has 3 shells:

the first shell ($n = 1$) contains 2 electrons

the second shell ($n = 2$) contains 8 electrons

the third shell ($n = 3$) contains 1 electron.

The equation representing the second ionisation energy of sodium is shown below.

$$Na^+(g) \longrightarrow Na^{2+}(g) + e^-$$ 2nd ionisation energy = +4563 kJ mol^{-1}

Sodium, with 11 electrons, has 11 successive ionisation energies, shown as the graph below. Successive ionisation energies provide evidence for the different energy levels of the principal quantum numbers.

These **2** electrons are in the 1st shell ($n = 1$), closest to the nucleus. These electrons experience very **strong** attractive forces from the nucleus.

This **1** electron is in the 3rd shell ($n = 3$), furthest from the nucleus. This electron experiences comparatively **weak** attractive forces from the nucleus.

ionisation energy /kJ mol^{-1}

ionisation number

These **8** electrons are in the 2nd shell ($n = 2$). These electrons experience reasonably strong attractive forces from the nucleus.

Trends in first ionisation energies

AQA	M1	OCR	M1
EDEXCEL	M1	SALTERS	M2
CCEA	M1	WJEC	M2

The graph below shows the variation of first ionisation energy with increasing atomic number from hydrogen to calcium:

Each 'peak' marks a noble gas (He, Ne, Ar) at the end of each period.

Each 'trough' marks an alkali metal (Li, Na, K) at the start of each period

first ionisation energy

atomic number

Across a period, increased nuclear charge is most important.

Down a group, increased distance and shielding are most important.

The graph shows:

- a general **increase** in first ionisation energy across a period (see H→He; Li→Ne; Na→Ar).
  This results from the **increase** in nuclear charge as electrons are added to the **same shell** across each period.
- a sharp **decrease** in first ionisation energy between the end of one period and the start of the next period (see He→Li; Ne→Na; Ar→K).
  This reflects the addition of a new outer shell with the resulting increase in **distance** and **shielding**.
- a **decrease** in first ionisation energy down a group (see He→Ne→Ar and other groups).
  This reflects the presence of **extra shells** down the group.

(handwritten annotations):
- Strong attraction of e + nucleus
- Same shielding
- high nuclear charge
- increases
- decreases, less attraction, more shielding, more charge
- Mg–Al decrease, Al – in 3p sub shell — 3p has more shielding, more energy, less attraction

The reasons for the general trend in ionisation energies are similar to those explaining the trend in atomic radii (page 73).

Evidence for sub-shells

AQA   M1
EDEXCEL   M1
CCEA   M1

The variation in first ionisation energies across a period in the Periodic Table provides evidence for the existence of sub-shells. The graph below shows this variation across Period 2 (Li → Ne)

Across the period, the graph shows:

• a rise from lithium to beryllium,
• a fall to boron followed by a rise to nitrogen,
• a fall to oxygen followed by a rise to neon.

Comparing beryllium and boron

The fall in 1st ionisation energy from beryllium to boron marks the start of filling the 2p sub-shell which is at higher energy than the 2s sub-shell.

Beryllium versus boron: higher energy level.

• Beryllium's outermost electron is in the 2s sub-shell.
• Boron's outermost electron is in the 2p sub-shell.
• Boron's outermost electron is easier to remove because the 2p sub-shell has a **higher energy level** than the 2s sub-shell.

- 3p subshell has more shielding, more energy than 3s.
  ∴ Al has lower Ionisation energy than Mg.

- Sulphur has paired electrons. which is easier to remove than non paired electrons in phosphrus. ∴ phosphrus
is greater in energy more electron repulsion.
   P8 > S4 > Cl2.

Comparing nitrogen and oxygen

The fall in 1st ionisation energy from oxygen to nitrogen marks the start of electron pairing in the p-orbitals of the 2p sub-shell.

easier to remove

Nitrogen versus oxygen: electron pairing.

N > O

- Both nitrogen and oxygen have their outer electrons in the same 2p sub-shell with the same energy level.
- Oxygen has one 2p orbital with an electron pair.
- Nitrogen, with one electron in each 2p orbital, has unpaired 2p electrons only.
- Because of **electron repulsion**, it is easier to remove one of oxygen's paired 2p electrons.

Nitrogen
2p sub-shell half-full
1 electron in each 2p orbital
parallel spins at right angles

Oxygen
2p electrons start to pair
Paired electrons repel

Atomic spectra of hydrogen

SALTERS	M1
WJEC	M1
CCEA	M1

Emission and absorption

The emission and absorption spectra of atomic hydrogen are the result of electron transitions between atomic energy levels.

Each element has its own unique set of lines in its atomic spectra. This allows elements to be identified. We know that the Sun contains hydrogen and helium gases because their atomic spectra have been observed.

- Energy is absorbed, exciting an electron from its ground state to a higher energy level.
- The electron returns to lower energy levels, emitting energy.

An **absorption** spectrum shows as black lines on a coloured or bright background.

- The **black lines** show that **radiation is absorbed** as electrons are excited from a **lower to higher** energy level.

An **emission** spectrum shows as bright or coloured lines on a black background.

- The **bright lines** show that **radiation is emitted** as electrons move from **higher to lower** energy levels.

Energy and frequency

The energy emitted or absorbed is proportional to the frequency of a line produced in the spectrum.

frequency, f, is related to the speed of light, c, and wavelength, λ:

$f = c/\lambda$

energy change, $\Delta E = hf$

h = the Planck constant; f = frequency

The hydrogen spectrum

The emission spectrum of atomic hydrogen has several series of lines.

- The lines in a series get closer together with increasing frequency.
- Eventually the lines converge as the energy levels also converge.

This provides evidence for discrete (separate and distinct) energy levels.

- Transitions from a higher energy level to the $n = 1$ energy level (the ground state of hydrogen) are in the ultraviolet region of the spectrum. This is the Lyman series.
- Transitions from a higher energy level to the $n = 2$ energy level are in the visible region of the spectrum. This is the Balmer series.

> Emission and absorption spectra both have lines in the same position for a given element:
>
> sets of lines represent transitions to or from a particular level
>
> the lines become closer at higher frequencies.

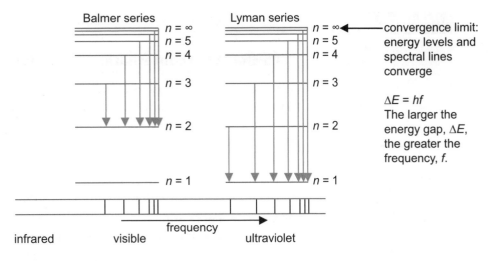

convergence limit: energy levels and spectral lines converge

$\Delta E = hf$
The larger the energy gap, ΔE, the greater the frequency, f.

Calculating the ionisation energy

The ionisation energy of hydrogen can be calculated from the frequency of the convergence limit in the Lyman series.

In the Lyman series, the lines converge at $\lambda = 91.1$ nm.

$f = c/\lambda = 3.00 \times 10^8/91.1 \times 10^{-9} = 3.29 \times 10^{15}$ s^{-1}

$\Delta E = hf = 6.62 \times 10^{-34} \times 3.29 \times 10^{15} = 2.18 \times 10^{-18}$ J $= 2.18 \times 10^{-21}$ kJ

The first ionisation energy $= 6.02 \times 10^{23} \times 2.18 \times 10^{-21} = 1310$ kJ mol^{-1}.

> $c = 3.00 \times 10^8$ m s^{-1}
>
> $h = 6.62 \times 10^{-34}$ J s
>
> The calculation gives the energy per H atom. We need to multiply by the Avogadro constant (6.02×10^{23} mol^{-1}) to get the energy per mole of H atoms.

Progress check

1 State and explain the general trend in 1st ionisation energy across a period.

2 Explain why the 1st ionisation energy falls slightly:
  (a) between beryllium and boron
  (b) between nitrogen and oxygen.

1 The 1st ionisation energy decreases across a period. Across a period the number of protons increases and electrons are being added to the same shell. Therefore attraction from the nucleus increases.
2 (a) In B, the highest energy electron is in the 2p sub-shell, higher in energy than the 2s sub-shell in Be.
  (b) In O, 2 electrons pair up in a 2p sub-shell. In N, the 2p sub-shell is half filled with a single electron in each orbital. The paired electrons in O repel one another slightly making it easier to remove an electron.

1.4 Atomic mass

After studying this section you should be able to:

- *define relative masses, based on the ^{12}C scale*
- *describe the basic principles of the mass spectrometer*
- *calculate the relative atomic mass of an element from its isotopic abundance or mass spectrum*
- *calculate the relative molecular mass of a compound from relative atomic masses*

LEARNING SUMMARY

Relative atomic mass

AQA	M1	OCR	M1
EDEXCEL	M1	SALTERS	M1
CCEA	M1	WJEC	M1

Carbon-12: an international standard

The mass of an atom is too small to be measured on even the most sensitive balance. Chemists use **relative** masses to compare the atomic masses of different elements. First, we must have an atom with which to compare other atoms and an atom of the **carbon-12 isotope** is chosen as the international standard for the measurement of atomic mass.

- The mass of an atom of carbon-12 is exactly 12 unified atomic mass units (u).
- The mass of one-twelfth of an atom of carbon-12 is exactly 1 u.

All atoms in a pure isotope have the same atomic structure and the same mass. An isotope's *relative* mass is found by comparison with carbon-12.

> **Relative isotopic mass** is the mass of an atom of an isotope compared with one-twelfth the mass of an atom of carbon-12.
>
> **KEY POINT**

Unified atomic mass unit (u)

1 unified atomic mass unit (u) is the mass of one-twelfth the mass of an atom of the carbon-12 isotope. This provides the base measurement for atomic masses:

$1\ u = 1.661 \times 10^{-23}\ g$

In most chemistry work, it is reasonable to:
- neglect the tiny contribution to atomic mass from electrons
- take both the mass of a proton and a neutron as 1 u.

Relative isotopic mass is then simply the mass number of the isotope, e.g. the relative isotopic mass of ^{16}O is 16.

Relative atomic mass

Most elements consist of a mixture of isotopes, each with a different mass number. To work out the relative atomic mass of an element we must find the **weighted average mass** of the isotopes present from:

- the natural abundances of the isotopes
- the relative isotopic masses of the isotopes.

> **Relative atomic mass**, A_r, is the weighted average mass of an atom of an element compared with one-twelfth of the mass of an atom of carbon-12.
>
> **KEY POINT**

Note that a relative mass has no units. It is simply a ratio of masses and any mass units will cancel.

Examples of relative atomic masses:

H: 1.008
Cl: 35.45
Pb: 207.19

Calculating a relative atomic mass

Naturally occurring chlorine consists of 75% ^{35}Cl and 25% ^{37}Cl.

$\frac{75}{100}$ of chlorine is the ^{35}Cl isotope;

$\frac{25}{100}$ of chlorine is the ^{37}Cl isotope.

The relative atomic mass of chlorine $= \frac{75}{100} \times 35 + \frac{25}{100} \times 37 = 35.5$

In most chemistry work, the relative isotopic mass can be taken as a whole number.

Measuring relative atomic masses

A relative atomic mass can be determined using a **mass spectrometer**.

Remember VIADD:

Vaporisation
Ionisation
Acceleration
Deflection
Detection

- A sample of the element is placed into the mass spectrometer and vaporised.
- The sample is bombarded with electrons forming positive ions.
- The positive ions are accelerated using an electric field.
- The positive ions are deflected using a magnetic field.
- Ions of lighter isotopes are deflected more than ions of heavier isotopes. This separates different isotopes.
- The ions are detected to produce a **mass spectrum**.

You can follow the stages in the diagram below:

Notice that ions are detected in the mass spectrometer. The ions are formed following electron bombardment.

(Salters only) Time-of-flight mass spectrometry uses a different method to identify ions.

The ions are accelerated by an electric field of known strength.

The time taken for the ions to reach a detector is measured.

Lighter ions travel faster than heavier ions.

An ion is identified from the time it takes to reach the detector.

Relative atomic mass from a mass spectrum

A mass spectrum provides:

- the relative isotopic masses of the isotopes in an element
- isotopic abundances.

The diagram below shows how a relative atomic mass can be calculated from a mass spectrum.

Mass spectrum of a copper sample

Mass spectrometers have been sent into space to identify elements, e.g. Mars space probe. On Earth, mass spectrometers are used in environmental monitoring and for forensic analysis.

ion	relative mass	percentage abundance
$^{63}Cu^+$	63	70%
$^{65}Cu^+$	65	30%

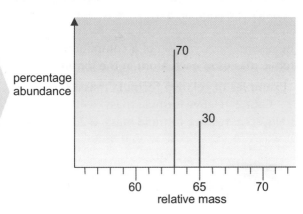

The relative atomic mass of copper

$$= \frac{70}{100} \times 63 + \frac{30}{100} \times 65$$

— long.

$$= 63.6$$

$$\frac{(70 \times 63) + (30 \times 65)}{100} = 63.6$$

More relative masses

AQA	M1	OCR	M1
EDEXCEL	M1	SALTERS	M1
CCEA	M1	WJEC	M1

Relative masses can be used to compare the masses of any chemical species. Much of matter is made up of compounds.

Relative molecular mass

You will find out more about chemical bonding and structure in Chapter 2, p.51.

A simple molecule of Cl_2, H_2O or CO_2 comprises small groups of atoms, held together by chemical bonds.

> **KEY POINT**
> Relative molecular mass, M_r, is the weighted average mass of a molecule of a compound compared with one-twelfth the mass of an atom of carbon-12.

Mass spectrometry can be used with compounds to determine the relative molecular mass.

The *relative molecular mass* of a compound is found by adding together the relative atomic masses of each atom in a molecule.

Examples of relative molecular masses
Cl_2: $M_r = 35.5 \times 2 = 71.0$;
H_2O: $M_r = 1.0 \times 2 + 16.0 = 18.0$

For giant structures, the **formula unit** of the compound is the simplest ratio of atoms or ions in the structure.

Some compounds, such as sand SiO_2, exist as 'giant' molecules comprising hundreds of thousands of atoms bonded together. Each molecule in these structures is the size of each crystal and, in a sense, the whole crystal is a molecule.

- Each 'molecule' of SiO_2 has twice as many oxygen atoms as silicon atoms.

Many compounds, such as common salt NaCl, are made up of ions and not molecules. These ionic compounds do not have single molecules and form giant structures of oppositely charged ions.

- NaCl has the same number of sodium ions as chloride ions.

Relative formula mass can be used to describe the relative mass of any compound.

Although relative molecular mass is often used with these compounds, a better term to use is relative formula mass.

> **KEY POINT**
> Relative formula mass is the average mass of the formula unit of a compound compared with one-twelfth the mass of an atom of carbon-12.

The *relative formula mass* of a compound is found by adding together the relative atomic masses of each atom in the formula.

Examples of relative formula masses
$CaBr_2$: relative formula mass $= 40.1 + 79.9 \times 2 = 199.9$;
Na_3PO_4: relative formula mass $= 23.0 \times 3 + 31.0 + 16.0 \times 4 = 164.0$

Progress check

a) $\dfrac{(60 \times 69) + (40 \times 71)}{100}$

$= 69.8$

b) $\dfrac{(92.18 \times 28) + (4.71 \times 29) + (3.11 \times 30)}{100} = 28.1$

1 Calculate the relative atomic mass, A_r, of the following elements from their isotopic abundances:
(a) Gallium containing 60% ^{69}Ga and 40% ^{71}Ga.
(b) Silicon containing 92.18% ^{28}Si, 4.71% ^{29}Si and 3.11% ^{30}Si.

2 Use A_r values from the Periodic Table to calculate the relative formula mass of:
(a) CO_2
(b) NH_3
(c) Fe_2O_3
(d) $C_6H_{12}O_6$
(e) $Pb(NO_3)_2$.

2 (a) 44.0
(b) 17.0
(c) 159.6
(d) 180.0
(e) 331.2.

1 (a) Gallium: 69.8
(b) Silicon: 28.11.

1.5 The mole

After studying this section you should be able to:

- *understand the concept of the mole as the amount of substance*
- *calculate molar quantities using masses, solutions and gas volumes*

The mole

AQA	M1	OCR	M1
EDEXCEL	M1	SALTERS	M1
CCEA	M1	WJEC	M1

Counting and weighing atoms

To make a compound such as H_2O, it would be useful to be able to count out atoms. For H_2O, we would need twice as many hydrogen atoms as oxygen atoms. Atoms are far too small to be counted individually but they can be counted using the idea of relative mass:

element	H	C	O
relative mass	1	12	16

For 1 atom of H, C and O, the relative masses are in the ratio 1 : 12 : 16.
For 10 atoms of each, the ratio is 10 : 120 : 160 which is still 1 : 12 : 16.

> Provided we have the same number of atoms of each, the masses are in the ratio of the relative masses.

By measuring masses, chemists are able to actually count atoms.
For 1 g of H, 12 g of C and 16 g of O, the masses are still in the ratio 1 : 12 : 16.
1 g of H contains the same number of atoms as 12 g of C and 16 g of O.

Amount of substance

Chemists use a quantity called *amount of substance* for counting atoms.
Amount of substance is:
- given the symbol n
- measured using a unit called the **mole** (abbreviated as **mol**).

As with relative masses, amount of substance uses the carbon-12 isotope as the standard:
The amount n of carbon-12 atoms in 12 g of ^{12}C is 1 *mol*.

The actual **number** of atoms in 1 mol of carbon-12 is 6.02×10^{23}.

This can be expressed by saying:

'There are 6.02×10^{23} atoms per mole of carbon-12 atoms (or 6.02×10^{23} mol^{-1})'.

6.02×10^{23} mol^{-1} is a constant known as the **Avogadro constant**, *L*, or N_A.

> Note that carbon-12 is the **substance** here. So **amount of** *substance* translates to **amount of carbon-12.**

> This is a very powerful and useful idea. To make a water molecule from hydrogen and oxygen, we cannot **count** out the atoms but we can **weigh** them out in the correct ratio. To make H_2O, we need 2 atoms of hydrogen for every 1 atom of oxygen, which we can get by weighing out 2 g of hydrogen and 16 g of oxygen.
>
> **2 g of H contains twice as many atoms as 16 g of O.**
> **2 g of H atoms contains 2 mol of H atoms.**
> **16 g of O atoms contains 1 mol of O atoms.**

KEY POINT

Particle
A single particle of a substance may refer to an atom, a molecule, an ion, an electron or to any identifiable particle. Chemists refer to a collection of particles as a **chemical species**.

Amount of substance is not restricted just to atoms. We can have an amount of any chemical species: 'amount of atoms', 'amount of molecules', 'amount of ions', etc.

> **KEY POINT**
> A **mole** is the amount of substance that contains as many single particles as there are atoms in exactly 12 g of the carbon-12 isotope.

Using the definition of the mole above, this means that:

- 1 mole of hydrogen atoms, H, contains 6.02×10^{23} hydrogen atoms, H.
- 1 mole of oxygen molecules, O_2, contains 6.02×10^{23} oxygen molecules, O_2.
- 1 mole of electrons, e^-, contains 6.02×10^{23} electrons, e^-.

It is important to always refer to what the particle is.

For example:

'1 mole of oxygen' could refer to oxygen atoms, O, or to oxygen molecules, O_2.

It is always safest to quote the formula to which the amount of substance refers.

Moles from masses

AQA	M1	OCR	M1
EDEXCEL	M1	SALTERS	M1
CCEA	M1	WJEC	M1

Molar mass, *M*

> **KEY POINT**
> Molar mass, *M*, is the mass per mole of a substance.
> The units of molar mass are g mol⁻¹.

For the atoms of an element, the molar mass is simply the relative atomic mass in g mol⁻¹.

- Molar mass of Mg = 24.3 g mol⁻¹
- This simply means that each mole of Mg atoms has a mass of 24.3 g.

For a compound, the molar mass is found by adding together the relative atomic masses of each atom in the formula:

- Molar mass of CO_2 = 12 + (16 x 2) = 44 g mol⁻¹
- 1 mole of CO_2 has a mass of 44 g.

Molar mass is an extremely useful term that can be applied to any chemical species.

Examples of molar masses
H   1 g mol⁻¹
H_2O 18 g mol⁻¹
CO_2 44 g mol⁻¹
HNO_3 63 g mol⁻¹

> **KEY POINT**
> Amount of substance *n*, mass and molar mass are linked by the expression below:
> $$n = \frac{mass\ (in\ g)}{molar\ mass}$$

In 24 g of C, *amount of C* = $\frac{mass\ (in\ g)}{molar\ mass}$ = $\frac{24}{12}$ = 2 mol

In 11 g of CO_2, *amount of CO_2* = $\frac{mass\ (in\ g)}{molar\ mass}$ = $\frac{11}{44}$ = 0.25 mol

Progress check

1 (a) What is the amount of each substance (in mol) in the following:
   (i) 124 g P; (ii) 64 g O; (iii) 64 g O_2; (iv) 10.01 g $CaCO_3$;
   (v) 3.04 g Cr_2O_3?
   (b) What is the mass of each substance (in g) in the following:
   (i) 0.1 mol Na; (ii) 2.5 mol NO_2; (iii) 0.3 mol Na_2SO_4;
   (iv) 0.05 mol H_2SO_4; (v) 0.025 mol Ag_2CO_3?

1 (a) (i) 4 mol; (ii) 4 mol; (iii) 2 mol; (iv) 0.1 mol; (v) 0.02 mol.
(b) (i) 2.3 g; (ii) 115 g; (iii) 42.63 g; (iv) 4.905 g; (v) 6.895 g.

Moles from solutions

AQA	M1	OCR	M1
EDEXCEL	M1	SALTERS	M2
CCEA	M1	WJEC	M1

> The concentration of a solution is the amount of solute, in mol, dissolved in each dm³ (1000 cm³) of solution.

'Dilute' solutions contain few moles of solute in a volume.

If the volume of the solution is in dm³:

$$n = c \times V \text{ (in dm}^3\text{)}$$

$n =$ amount of substance, in mol.

$c =$ concentration of solution, in mol dm⁻³.

$V =$ volume of solution, in dm³.

Dividing a volume measured in cm³ by 1000 converts the volume automatically to dm³.

It is more convenient to measure smaller volumes of solutions in cm³ and the expression becomes:

$$n = c \times \frac{V \text{ (in cm}^3\text{)}}{1000}$$

Molarity

Molarity refers to the concentration in mol dm⁻³.

Thus 2 mol dm³ and 2 Molar (or 2M) mean the same: 2 moles of solute in 1 dm³ of solution.

Example

What is the amount of NaCl (in mol) in 25.0 cm³ of an aqueous solution of concentration 2.00 mol dm⁻³?

$$n(\text{NaCl}) = c \times \frac{V}{1000} = 2.00 \times \frac{25.0}{1000} = 0.0500 \text{ mol}$$

Standard solutions

Concentrations are sometimes referred to in g dm⁻³.

The Na_2CO_3 here has a **molar** concentration of 0.100 mol dm⁻³ and a **mass** concentration of 10.6 g dm⁻³ of Na_2CO_3.

Chemists often need to prepare **standard solutions** with an exact concentration. Using an understanding of the mole, the mass required to prepare such a solution can easily be worked out.

Example

Find the mass of sodium carbonate required to prepare 250 cm³ of a 0.100 mol dm⁻³ solution.

Find the amount of Na_2CO_3 (in mol) required in solution:

Very dilute solutions are measured in parts per million (ppm), e.g. ppm is 1 milligram (0.001 g) in 1 dm³ of solution.

$$\text{amount, } n, \text{ of } Na_2CO_3 = c \times \frac{V}{1000} = 0.100 \times \frac{250}{1000} = 0.0250 \text{ mol}$$

Convert moles to grammes

molar mass of Na_2CO_3 = 23.0 × 2 + 12.0 + 16.0 × 3 = 106.0 g mol⁻¹

$$n = \frac{\text{mass}}{\text{molar mass}} \quad \therefore \text{ mass} = n \times \text{molar mass}$$

mass of Na_2CO_3 required = 0.0250 × 106.0 = 2.65 g

Progress check

1 (a) What is the amount of each substance, in mol, in:
      (i) 250 cm³ of a 1.00 mol dm⁻³ solution;
      (ii) 10 cm³ of a 2.0 mol dm⁻³ solution?

   (b) Find the concentration, in mol dm⁻³, for:
      (i) 4 mol in 2 dm³ of solution;
      (ii) 0.0100 mol in 100 cm³ of solution.

   (c) Find the concentration, in g dm⁻³, for:
      (i) 2 mol of NaOH in 4 dm³ of solution;
      (ii) 0.500 mol of HNO_3 in 200 cm³ of solution.

1 (a) (i) 0.25 mol; (ii) 0.02 mol.
   (b) (i) 2 mol dm⁻³; (ii) 0.1 mol dm⁻³.
   (c) (i) 20 g dm⁻³; (ii) 157.5 g dm⁻³.

Moles from gas volumes

AQA	M1	OCR	M1
EDEXCEL	M1	SALTERS	M2
CCEA	M2	WJEC	M1

This is sometimes summarised by Avogadro's hypothesis:

'Equal volumes of gases contain the same number of molecules under the same conditions of temperature and pressure.'

For a gas, the amount of gas molecules (in mol) is most conveniently obtained by measuring the gas volume.

Provided that the pressure and temperature are the same, equal volumes of gases contain the same number of molecules.

This means that it does not matter which gas is being measured. By measuring the volume, we are indirectly also counting the number of molecules.

> **KEY POINT**
>
> At room temperature and pressure (r.t.p), 298 K (25°C) and 100 kPa, 1 mol of a gas occupies approximately 24 dm³ = 24 000 cm³.

At r.t.p., if the volume of the gas is in **dm³**: $n = \dfrac{V \text{ (in dm}^3)}{24}$

if the volume of the gas is in **cm³**: $n = \dfrac{V \text{ (in cm}^3)}{24000}$

Example

How many moles of gas molecules are in 480 cm³ of a gas at r.t.p.?

$$n = \frac{V \text{ (in cm}^3)}{24000} = \frac{480}{24000} = 0.0200 \text{ mol of gas molecules.}$$

The Ideal Gas Equation (AQA only)

The conditions will not always be room temperature and pressure. A gas volume depends on temperature and pressure.

The Ideal Gas Equation can be used to find the amount of gas molecules n (in mol) in a volume V (in m³) at any temperature T (in K) and pressure p (in Pa).

The Ideal Gas Equation: $pV = nRT$ ($R = 8.31$ J K⁻¹ mol⁻¹)

Before using $pV = nRT$, you must remember to convert any values to Pa, K and m³.

Conversion rules

cm³ to m³	× 10⁻⁶
dm³ to m³	× 10⁻³
°C to K	+ 273
kPa to Pa	× 10³

Note that M (molar mass) has units: g mol⁻¹

M_r (relative molecular mass) has no units.

Example

A 0.215 g sample of a volatile liquid, **X**, produces 77.5 cm³ of gas at 100°C and 100 kPa. Calculate the relative molecular mass of **X**.

Use The Ideal Gas Equation:

$$pV = nRT \quad \therefore n = \frac{pV}{RT}$$

$$\therefore n = \frac{1.00 \times 10^5 \times 7.75 \times 10^{-5}}{8.31 \times 373} = 0.00250 \text{ mol}$$

Find the molar mass:

$$n = \frac{\text{mass}}{\text{molar mass}}$$

$$\therefore \text{molar mass} = \frac{\text{mass}}{n} = \frac{0.215}{0.00250} = 86.0 \text{ g mol}^{-1}$$

$$\therefore \text{Relative molecular mass } M_r = 86.0$$

Progress check

1 (a) What is the volume, at room temperature and pressure (r.t.p.) of
   (i) 44 g CO_2(g); (ii) 7 g N_2(g); (iii) 5.1 g NH_3(g)?
   (b) What is the mass, at r.t.p. of
   (i) 1.2 dm³ O_2; (ii) 720 cm³ CO_2(g); (iii) 48 cm³ CH_4(g)?

1 (a) (i) 24 dm³; (ii) 6 dm³; (iii) 7.2 dm³.
(b) (i) 1.6 g; (ii) 1.32 g; (iii) 0.032 g.

1.6 Formulae, equations and reacting quantities

After studying this section you should be able to:

- *understand what is meant by empirical and molecular formula*
- *calculate the empirical and molecular formula of a substance*
- *write balanced equations*
- *use chemical equations to calculate reacting masses and reacting volumes (and vice versa)*
- *perform calculations involving volumes and concentrations of solutions in simple acid-base titrations and back titrations*

LEARNING SUMMARY

Types of chemical formula

AQA	M1	OCR	M1
EDEXCEL	M1	SALTERS	M1
CCEA	M2	WJEC	M1

Empirical and molecular formula

The **empirical formula** of ethane is CH_3.

Ethane has 1 carbon atom for each of 3 hydrogen atoms.

The molecular formula of ethane is C_2H_6

Each molecule of ethane contains 2 carbon atoms and 6 hydrogen atoms.

> **KEY POINT**
>
> **Empirical formula** represents the simplest, whole-number ratio of atoms of each element in a compound.
>
> **Molecular formula** represents the actual number of atoms of each element in a molecule of a compound.

Formula determination

A formula can be calculated from experimental results using the Mole Concept.

Example 1

Find the empirical formula of a compound formed when 6.75 g of aluminium reacts with 26.63 g of chlorine. [A_r: Al, 27.0; Cl, 35.5.]

Find the molar ratio of atoms:

$$
\begin{array}{ccc}
Al & : & Cl \\
= \dfrac{6.75}{27.0} & : & \dfrac{26.63}{35.5} \\
0.25 & : & 0.75
\end{array}
$$

Divide by smallest number (0.25):   1   :   3

∴ Empirical formula = $AlCl_3$

Example 2

Find the molecular formula of a compound containing carbon, hydrogen and oxygen only with the composition by mass of carbon: 40.0%; hydrogen: 6.7%. [$M_r = 180$. A_r: H, 1.00; C, 12.0; O, 16.0.]

Analysis often doesn't give the oxygen content directly but it can easily be calculated.

Percentage of oxygen in compound = 100 − (40.0 + 6.7) = 53.3%
100.0 g of the compound contains 40.0 g C, 6.7 g H and 53.3 g O.

Find the molar ratio of atoms:

$$
\begin{array}{ccccc}
C & : & H & : & O \\
= \dfrac{40.0}{12.0} & : & \dfrac{6.7}{1.00} & : & \dfrac{53.3}{16.0} \\
= 3.33 & : & 6.7 & : & 3.33
\end{array}
$$

Divide by smallest number (3.33):   1   :   2   :   1

∴ Empirical formula = CH_2O

Relate to molecular mass

Each CH_2O unit has a relative mass of 12 + (1 × 2) + 16 = 30
M_r, of the compound is 180 containing 180/30 = 6 CH_2O units
∴ molecular formula is $C_6H_{12}O_6$

Progress check

1 Find the empirical formula of the following:
   (a) sodium oxide (2.3 g of sodium reacts to form 3.1 g of sodium oxide).
   (b) iron oxide (11.16 g of iron reacts to form 15.96 g of iron oxide).

2 Use the following percentage compositions by mass to find the empirical formula of the compound.
   (a) A compound of sulfur and oxygen. [S, 40.0%; O, 60.0%]
   (b) A compound of iron, sulfur and oxygen. [Fe, 36.8%; S, 21.1%; O, 42.1%]

3 2.8 g of a compound of carbon and hydrogen is formed when 2.4 g of carbon combines with hydrogen [M_r: 56]. What is the empirical and molecular formula of the compound formed?

1 (a) Na_2O; (b) Fe_2O_3.
2 (a) SO_3; (b) $FeSO_4$.
3 CH_2; C_4H_8.

Equations

Chemical reactions involve the **rearrangement** of atoms and ions.

Chemical equations provide two types of information about a reaction:

- qualitative – *which* atoms or ions are rearranging;
- quantitative – *how* many atoms or ions are rearranging.

Balancing equations

The formula of each substance shows the chemicals involved in the reaction.

State symbols can be added to show the physical states of each species under the conditions of the reaction.

> *State* symbols give the physical states of each species in a reaction:
>
> gaseous state, (g);
> liquid state, (l);
> solid state, (s);
> aqueous solution, (aq).

The **qualitative** equation for the reaction of hydrogen with oxygen is shown below:

$$\text{hydrogen} + \text{oxygen} \longrightarrow \text{water}$$
$$H_2(g) + O_2(g) \longrightarrow H_2O(l)$$

In this equation, hydrogen is balanced, oxygen is **not** balanced.

	$H_2(g)$	+	$O_2(g)$	$\longrightarrow$	$H_2O(l)$	
hydrogen	2				2	✓
oxygen			2		1	✗

The equation must be **balanced** to give the **same number** of particles of each element on each side of the equation.

Balancing an equation often takes several stages.

- Oxygen can be balanced by placing a '2' in front of H_2O.
- However, this unbalances hydrogen:

	$H_2(g)$	+	$O_2(g)$	$\longrightarrow$	2 $H_2O(l)$	
hydrogen	2				4	✗
oxygen			2		2	✓

- H can be re-balanced by placing a '2' in front of H_2.
- The equation is now balanced.

> Notice that no formula has been changed. You balance an equation by adding numbers in front of a formula only.

	2 $H_2(g)$	+	$O_2(g)$	$\longrightarrow$	2 $H_2O(l)$	
hydrogen	4				4	✓
oxygen			2		2	✓

More balancing numbers.

3 Na_2SO_4 means
Na: $3 \times 2 = 6$;
S: $3 \times 1 = 3$;
O: $3 \times 4 = 12$.

With brackets, the subscript applies to everything in the preceding bracket:

2 $Ca(NO_3)_2$ means
Ca: $2 \times 1 = 2$;
N: $2 \times (1 \times 2) = 4$;
O: $2 \times (3 \times 2) = 12$.

> **KEY POINT**
>
> In a formula, a subscript applies **only** to the symbol immediately preceding,
>
> e.g. NO_2 comprises **1** N and **2** O.
>
> When balancing an equation, you must **not** change any formula. You can only add a balancing number in front of a formula.
>
> The balancing number multiplies everything in the formula by the balancing number,
>
> e.g. $2AlCl_3$ comprises Al: $2 \times 1 = 2$; Cl: $2 \times 3 = 6$.

Progress check

Balance the following equations:
(a) $Li(s) + O_2(g) \longrightarrow Li_2O(s)$
(b) $CH_4(g) + O_2(g) \longrightarrow CO_2(g) + H_2O(l)$
(c) $Al(s) + HCl(aq) \longrightarrow AlCl_3(aq) + H_2(g)$

(a) $4Li(s) + O_2(g) \longrightarrow 2Li_2O(s)$
(b) $CH_4(g) + 2O_2(g) \longrightarrow CO_2(g) + 2H_2O(l)$
(c) $2Al(s) + 6HCl(aq) \longrightarrow 2AlCl_3(aq) + 3H_2(g)$.

Reacting quantities

An understanding of molar reacting quantities provides chemists with the required recipes to make quantities of chemicals to order.

The balancing numbers give the ratio of the **amount** of each substance, in mol. Using the example above:

equation	$2 H_2(g)$	$+$	$O_2(g)$	$\longrightarrow$	$2 H_2O(l)$
moles	2 mol		1 mol	$\longrightarrow$	2 mol

Chemists use this **quantitative** information to find:
- the *reacting quantities* required to prepare a particular quantity of a product
- the *quantities of products* formed by reacting particular quantities of reactants.

By working out the reacting quantities from a balanced chemical equation, quantities can be adjusted to take into account the required scale of preparation.

Example 1

These examples assume a 100% yield of products. See also percentage yield, page 123.

Calculate the masses of nitrogen and oxygen required to form 3 g of nitrogen oxide, NO. [A_r: N, 14; O, 16].

equation	$N_2(g)$	$+$	$O_2(g)$	$\longrightarrow$	$2 NO(g)$
moles	1 mol	$+$	1 mol	$\longrightarrow$	2 mol
reacting masses	(14×2) g	$+$	(16×2) g	$\longrightarrow$	$2 (14 + 16)$ g
	28 g	$+$	32 g	$\longrightarrow$	60 g
for 3 g NO(g), $\div 20$:	1.4 g	$+$	1.6 g	$\longrightarrow$	3 g

∴ 1.4 g of $N_2(g)$ react with 1.6 g of $O_2(g)$ to form 3 g NO(g).

Example 2

You must keep the proportions the same as the reacting quantities.

Any scaling must be applied to all reacting quantities.

Calculate the volume of oxygen, at r.t.p., formed by the decomposition of an aqueous solution containing 5 g of hydrogen peroxide, H_2O_2. [A_r: H, 1; O, 16.]

equation	$2 H_2O_2(aq)$	$\longrightarrow$	$2 H_2O(l)$	$+$	$O_2(g)$
moles	2 mol	$\longrightarrow$	2 mol	$+$	1 mol
reacting quantities	$2 [(1 \times 2) + (16 \times 2)]$ g	$\longrightarrow$			1×24 dm³
	68 g	$\longrightarrow$			24 dm³
for 1 g H_2O_2, $\div 68$,	1 g	$\longrightarrow$			$\frac{24}{68}$ dm³
for 5 g H_2O_2, $\times 5$,	5 g	$\longrightarrow$			$\frac{24}{68} \times 5$ dm³

∴ 5 g of H_2O_2 decomposes to form 1.8 dm³ of $O_2(g)$

Progress check

1 For the reaction: $S(s) + O_2(g) \longrightarrow SO_2(g)$
   (a) How many moles of S and O_2 react to form 0.5 mole of SO_2?
   (b) What mass of SO_2 is formed by reacting 64.2 g of S with O_2?
   (c) What volume of SO_2 forms when 0.963 g of S reacts with O_2 at r.t.p.?

2 Balance the equation: $Mg(s) + O_2(g) \longrightarrow MgO(s)$
   (a) What mass of MgO forms by burning 8.1 g of Mg?
   (b) What volume of O_2 at r.t.p. will react with this mass of Mg?
   (c) What masses of Mg and O_2 are needed to prepare 1 g of MgO?

2 (a) 13.4 g; (b) 4 dm³; (c) 0.6 g Mg, 0.4 g O₂.
1 (a) 0.5 mol S, 0.5 mol O₂; (b) 96.2 g; (c) 0.72 dm³.

Calculations in acid-base titrations

AQA	M1	OCR	M1
EDEXCEL	M2	SALTERS	M2
CCEA	M1	WJEC	M1

The acid is added from the burette to the alkali.

An indicator is required to show the 'end-point' when all alkali has reacted with the added acid.

The concentration and volume of NaOH are known.

Using the equation determine the number of moles of the second reagent

Calculations for acid-base titrations are best shown with an example:

In a titration, 25.0 cm³ of 0.100 mol dm⁻³ sodium hydroxide NaOH(aq) were found to react exactly with 20.80 cm³ of sulfuric acid, H_2SO_4(aq). Find the concentration of the sulfuric acid.

To solve the problem, we must use the following pieces of information:

- the balanced equation
$$2NaOH(aq) + H_2SO_4(aq) \longrightarrow Na_2SO_4(aq) + 2H_2O (l)$$
- the concentration c_1 and reacting volume V_1 of NaOH(aq)
- the concentration c_2 and reacting volume V_2 of H_2SO_4(aq).

From the titration results, the amount of NaOH (in mol) can be calculated:
$$\text{amount of NaOH} = c \times \frac{V}{1000} = 0.100 \times \frac{25.0}{1000} = 0.00250 \text{ mol}$$

From the equation, the amount of H_2SO_4 (in mol) can be determined:

$$2 \text{ NaOH (aq)} + H_2SO_4(aq) \longrightarrow Na_2SO_4(aq) + 2H_2O (l)$$
2 mol   1 mol *(balancing numbers)*

∴ 0.00250 mol NaOH reacts with 0.00125 mol H_2SO_4
amount of H_2SO_4 that reacted = 0.00125 mol

The concentration (in mol dm⁻³) of H_2SO_4 can be calculated by scaling to 1000 cm³:

20.80 cm³ H_2SO_4(aq) contains 0.00125 mol H_2SO_4

1 cm³ H_2SO_4(aq) contains $\frac{0.00125}{20.80}$ mol H_2SO_4

Work out the concentration of H_2SO_4(aq), in mol dm⁻³.

1 dm³ (1000 cm³) H_2SO_4(aq) contains $\frac{0.00125}{20.80} \times 1000 = 0.0601$ mol H_2SO_4

∴ concentration of H_2SO_4(aq) is 0.0601 mol dm⁻³

Back titrations (CCEA only)

A back titration is a technique that can be used to analyse a sample that is insoluble in water.

In a **back titration**:

- a sample is first reacted with an **excess** of a solution of which the volume and concentration are both known
- the excess amount added is determined by carrying out a titration.

From the results, we can find out information about the sample.

AS exam questions would help you through titrations such as this by asking each step in the calculation as a question.

Worked Example

75.0 cm³ of 0.500 mol dm⁻³ hydrochloric acid is added to 1.25 g of an impure sample of magnesium carbonate, $MgCO_3$.

The equation for the reaction between HCl and $MgCO_3$ is shown below:

equation 1 $2HCl(aq) + MgCO_3(s) \longrightarrow MgCl_2(aq) + H_2O(l) + CO_2(g)$

This is an excess of HCl:

- all of the $MgCO_3$ reacts
- some hydrochloric acid is left over unreacted.

The excess of HCl(aq) is titrated with 0.500 mol dm⁻³ sodium hydroxide and 21.60 cm³ is required for complete reaction.

The equation for the reaction is shown below:

equation 2 $NaOH(aq) + HCl(aq) \longrightarrow NaCl(aq) + H_2O(l)$

Calculate the percentage of $MgCO_3$ in the impure sample.

From the titration results, the amount (in mol) of NaOH can be calculated:

The concentration and volume of NaOH are known.

$$\text{amount of NaOH} = c \times \frac{V}{1000} = 0.500 \times \frac{21.60}{1000} = 0.0108 \text{ mol}$$

From equation 2, the amount (in mol) of HCl that reacts with the NaOH can be determined:

Using the equations, determine the number of moles of HCl

From equation 2,
1 mol NaOH reacts with 1 mol HCl
∴ 0.0108 mol NaOH reacts with 0.0108 mol HCl
This is the excess of HCl left over unreacted.

The amount (in mol) of HCl that reacted with the sample can now be determined:

At the start, 75.0 cm³ of 0.500 mol dm⁻³ hydrochloric acid is used.

The amount of HCl that reacted in equation 1 is now found.

$$\text{amount (in mol) of HCl used} = 0.500 \times \frac{75.0}{1000} = 0.0375 \text{ mol}$$

amount of HCl that reacts = original amount HCl – excess amount HCl
= 0.0375 – 0.0108 = 0.0267 mol

Calculate the mass of $MgCO_3$ in the rock sample:

Using the equations, determine the number of moles of $MgCO_3$.

From equation 1,
2 mol HCl reacts with 1 mol $MgCO_3$.
∴ 0.0267 mol HCl reacts with $\frac{0.0267}{2}$ = 0.01335 mol $MgCO_3$.
mass of $MgCO_3$ = 0.01335 × (24.3 + 12 + 16 × 3) = 1.125 g

Calculate the percentage by mass of $MgCO_3$ in the impure sample:

$$\text{Percentage of } MgCO_3 \text{ in impure sample} = \frac{\text{mass of } MgCO_3}{\text{mass of impure } MgCO_3} \times 100$$

$$= \frac{1.125}{1.25} \times 100 = 90.0\%$$

Progress check

1 25.0 cm³ of 0.500 mol dm⁻³ NaOH(aq) reacts with 23.2 cm³ of HNO_3(aq).

$$HNO_3(aq) + NaOH(aq) \longrightarrow NaNO_3(aq) + H_2O(l)$$

Find the concentration of the nitric acid, HNO_3.

2 25.0 cm³ of 0.100 mol dm⁻³ KOH(aq) reacts with 26.6 cm³ of H_2SO_4(aq).

$$H_2SO_4(aq) + 2KOH(aq) \longrightarrow K_2SO_4(aq) + 2H_2O(l)$$

Find the concentration of the sulfuric acid, H_2SO_4.

2 0.0470 mol dm⁻³.

1 0.539 mol dm⁻³.

1.7 Redox reactions

After studying this section you should be able to:

- *explain the terms reduction and oxidation in terms of electron transfer*
- *explain the terms oxidising agent and reducing agent*
- *apply the rules for assigning oxidation states*
- *identify changes in oxidation numbers from an equation*
- *construct an overall equation for a redox reaction from half-equations*

LEARNING SUMMARY

Oxidation and Reduction

AQA	M2	OCR	M1
EDEXCEL	M2	SALTERS	M2
CCEA	M1	WJEC	M2

Redox reactions

Oxidation and *reduction* were originally used for reactions involving oxygen.

> **Oxidation is the gain of oxygen.**
> **Reduction is the loss of oxygen.**
>
> **KEY POINT**

Nowadays, oxidation and reduction have a much broader definition in terms of electron transfer in a *redox* reaction (**red**uction and **ox**idation).

> **Reduction is the gain of electrons.**
> **Oxidation is the loss of electrons.**
>
> **KEY POINT**

Reduction and oxidation must take place together:

- if one species **gains** electrons
- another species **loses** the same number of electrons

OIL
RIG

Oxidation
Is
Loss of electrons
Reduction
Is
Gain of electrons

Half-equations

The formation of magnesium chloride from its elements is a redox reaction:

overall reaction $\qquad Mg + Cl_2 \longrightarrow MgCl_2$

The overall equation conceals the electron transfer that has taken place. This can be shown by writing half-equations:

electron transfer $\qquad Mg \longrightarrow Mg^{2+} + 2e^-$ oxidation (loss of electrons)
$\qquad\qquad\qquad\qquad Cl_2 + 2e^- \longrightarrow 2Cl^-$ reduction (gain of electrons)

Half-equations are useful to identify the species being oxidised or reduced. Notice that the number of electrons lost and gained must balance.

Oxidising and reducing agents

> An **oxidising agent accepts** electrons from another reactant.
> Non-metals are oxidising agents, e.g. F_2, Cl_2, O_2.
> A **reducing agent donates** electrons to another reactant.
> Metals are reducing agents, e.g. Na, Fe, Zn.
>
> **KEY POINT**

Non-metals are oxidising agents.
Metals are reducing agents.

In the example above:

- Mg is the reducing agent – it has *reduced* the Cl_2 to $2Cl^-$ by *adding* electrons
- Cl_2 is the oxidising agent – it has *oxidised* Mg to Mg^{2+} by *removing* electrons.

Using oxidation numbers

AQA	M2	OCR	M1
EDEXCEL	M2	SALTERS	M2
CCEA	M1	WJEC	M2

Chemists use the concept of **oxidation number** as a means of accounting for electrons.

The oxidation number of a species is assigned by applying a set of rules.

Oxidation number rules

Exceptions to rules

In compounds with fluorine and in peroxides, the oxidation number of oxygen is not –2 and must be calculated from other oxidation numbers.

In metal hydrides, the oxidation number of hydrogen is –1.

species	oxidation number	examples
uncombined element	0	C, zero; Na, zero; O_2, zero.
combined oxygen	–2	H_2O; CaO.
combined hydrogen	+1	NH_3, H_2S.
simple ion	charge on ion	Na^+, +1; Mg^{2+}, +2; Cl^-, –1
combined fluorine	–1	NaF, CaF_2.

> **KEY POINT**
>
> When applying oxidation numbers to elements, compounds and ions, the sum of the oxidation numbers must equal the overall charge.

Examples of applying oxidation numbers

Oxidation Number rules can be applied to any compound, whether ionic or covalent.

In CO_2, the overall charge is zero.

- There are **2** oxygen atoms, each with an oxidation number of –2, giving a total contribution of **–4**.
- The oxidation number of carbon must be **+4** to give the overall charge of zero.

In NO_3^-, the overall charge is 1–.

- There are **3** oxygen atoms, each with an oxidation number of –2, giving a total contribution of **–6**.
- The oxidation number of nitrogen must be **+5** to give the overall charge of –1.

Oxidation number applies to each atom in a species.

Notice how this has been shown in the examples.

In a compound:	CO_2
• the total of all the oxidation numbers is zero:	+4
	–2
	–2
	+4 –4
CO_2: overall charge = 0	0

In an ion:	NO_3^-
• the total of all the oxidation numbers is equal to the overall charge on the ion:	+5
	–2
	–2
	–2
	+5 –6
NO_3^-: overall charge = –1	–1

Sometimes an element can form compounds or ions in which its atoms can have different oxidation states. The oxidation number is included in the name as a Roman numeral.

For example, there are two nitrate ions:

nitrate(III), NO_2^- N: +3

nitrate(V), NO_3^- N: +5

Using oxidation number with equations

By applying oxidation numbers to an equation:

- the species being oxidised and reduced can be identified
- the number of electrons on both sides of the equation can be checked.

Oxidation numbers can be used to identify redox reactions in which electron loss and electron gain are not easy to see.

The sum of the oxidation numbers on both sides of a chemical equation must be the same.

	Cr_2O_3 (s)	+	2 Al (s)	$\longrightarrow$	Al_2O_3 (s)	+	2 Cr (s)
oxidation	+3 –2		0		+3 –2		0
numbers	+3 –2		0		+3 –2		0
	–2				–2		
sum of oxidation numbers:	0		0		0		0

In this reaction, the changes in oxidation number are:

Cr: $+3 \longrightarrow 0$ reduction   *oxidation number decreases*
Al: $0 \longrightarrow +3$ oxidation   *oxidation number increases*

> **KEY POINT**
>
> Oxidation is an **increase** in oxidation number.
> Reduction is a **decrease** in oxidation number.

Combining half-equations

An overall equation for a redox reaction can be constructed by combining the half-equations showing the transfer of electrons in the reaction.

Example: The reaction of Ag+ ions react with Zn metal.

electron gain: reduction

electron loss: oxidation

The half-equations are:

$$Ag^+(aq) + e^- \longrightarrow Ag(s)$$
$$Zn(s) \longrightarrow Zn^{2+}(aq) + 2e^-$$

To combine the half-equations in an overall equation, the number of electrons transferred must be the same in each half-equation. This will ensure that every electron lost by Zn(s) is gained by Ag+.

Balance the electrons: Ag+ reaction x 2.

- The Ag+ half-equation is multiplied by '2' to balance the electrons.
$$2Ag^+(aq) + 2e^- \longrightarrow 2Ag(s)$$

- The half-equations are now added:
$$2Ag^+(aq) + 2e^- + Zn(s) \longrightarrow 2Ag(s) + Zn^{2+}(aq) + 2e^-$$

Cancel the electrons to give the overall equation.

- Any species appearing on both sides are cancelled to give the overall equation.
$$2Ag^+(aq) + Zn(s) \longrightarrow 2Ag(s) + Zn^{2+}(aq)$$

- Finally, check that the oxidation numbers balance on either side of the equation:

	$2Ag^+(aq)$	+	$Zn(s)$	$\longrightarrow$	$2Ag(s)$	+	$Zn^{2+}(aq)$
oxidation numbers:	+1		0		0		+2
	+1				0		

oxidation number check: +2   +2

Progress check

1 What is the oxidation state of the elements in the following:
(a) Ag+; (b) F_2; (c) N^{3-}; (d) Fe; (e) MgF_2; (f) $NaClO_3$?

2 Write down the oxidation number of sulfur in the following:
(a) H_2S; (b) SO_2; (c) SO_4^{2-}; (d) SO_3^{2-}; (e) $Na_2S_2O_3$.

3 The following reaction is a redox process:
$Mg + 2HCl \longrightarrow MgCl_2 + H_2$
(a) Identify the changes in oxidation number.
(b) Which species is being oxidised and which is being reduced?
(c) Identify the oxidising agent and the reducing agent.

1 (a) +1; (b) 0; (c) −3; (d) 0; (e) Mg: +2, F: −1; (f) Na: +1, Cl: +5, O: −2.
2 (a) −2; (b) +4; (c) +6; (d) +4; (e) +2.
3 (a) Mg, 0 $\longrightarrow$ +2, H, +1 $\longrightarrow$ 0; (b) Mg oxidised, H reduced;
(c) Oxidising agent: HCl, Reducing agent: Mg.

1.8 Acids, bases and salts

After studying this section you should be able to:

- *know what is meant by acid, base, alkali and salt*
- *know what is meant by neutralisation*
- *describe the typical reactions of acids*
- *understand what is meant by hydrated, anhydrous and water of crystallisation*
- *know how to determine the water of crystallisation in a hydrated salt*

LEARNING SUMMARY

Acids, bases and alkalis

| OCR | M1 |
| WJEC | M1 |

Acids

A hydrogen ion, H^+ is simply a hydrogen atom with an electron removed.

Because a hydrogen atom comprises one proton and one electron, removing the electron leaves just a proton.

'H^+', 'Hydrogen ion' and 'proton' refer to the same thing.

The common acids are: H_2SO_4, HCl and HNO_3

> **KEY POINT**
>
> All acids release $H^+(aq)$ ions in aqueous solution.

Common acids include:

- strong acids: sulfuric acid, H_2SO_4; hydrochloric acid, HCl; nitric acid, HNO_3
- weak organic acids: ethanoic acid, CH_3COOH; methanoic acid, HCOOH

When added to water, HCl ionises and releases protons into the solution.

$$HCl(g) \quad + \quad aq \quad \longrightarrow \quad H^+(aq) \quad + \quad Cl^-(aq)$$

Bases

> **KEY POINT**
>
> All bases accept $H^+(aq)$ ions.

Common bases include:

- metal oxides: magnesium oxide, MgO; copper oxide, CuO
- metal hydroxides: sodium hydroxide, NaOH; iron(III) hydroxide, $Fe(OH)_3$
- ammonia and amines: ammonia, NH_3; methylamine, CH_3NH_2

Alkalis

> **KEY POINT**
>
> An alkali is a soluble base that dissolves in water, releasing $OH^-(aq)$ ions.

Common alkalis are: NaOH, KOH and NH_3

You need to learn these formulae as you will use these compounds very often.

There are only a few common alkalis:

- strong alkalis: potassium hydroxide, KOH; lithium hydroxide, LiOH
- weak alkalis: ammonia, NH_3

When added to water, the alkali NaOH ionises and releases hydroxide ions into the solution.

$$NaOH(s) \quad + \quad aq \quad \longrightarrow \quad Na^+(aq) \quad + \quad OH^-(aq)$$

Neutralisation

A base neutralises the H^+ of the acid to form water. The equation for neutralisation of an acid by an alkali is shown below:

$$H^+(aq) \quad + \quad OH^-(aq) \quad \longrightarrow \quad H_2O(l)$$

Salts

When acids react, they often form salts.

A salt is a compound formed from an acid, when a H^+ ion from the acid has been replaced by a metal ion or another positive ion, such as the ammonium ion, NH_4^+.

Examples of some common salts are shown below:

acid		salt	example
sulfuric acid	H_2SO_4	sulfate	sodium sulfate: Na_2SO_4
hydrochloric acid	HCl	chloride	sodium chloride: NaCl
nitric acid	HNO_3	nitrate	potassium nitrate: KNO_3

Typical reactions of acids

OCR ▶ M1

Acids react with carbonates, bases, alkalis and metals.

Neutralisation reactions

You will often come across these reactions of acids.

You need to learn how acids react with carbonates, bases, alkalis and metals.

Salts are always formed in solution from these typical acid reactions. The salt can be isolated by evaporation of water.

Acids react with carbonates, bases and alkalis in **neutralisation** reactions.

• The H^+ of the acid is neutralised to form water.

Acids react with carbonates, forming a salt, carbon dioxide gas and water:

$$2HCl(aq) + CaCO_3(s) \longrightarrow CaCl_2(aq) + CO_2(g) + H_2O(l)$$
acid + carbonate ⟶ salt + carbon dioxide + water

Acids react with bases, forming a salt and water:

$$2HCl(aq) + CaO(s) \longrightarrow CaCl_2(aq) + H_2O(l)$$
acid + base ⟶ salt + water

Acids react with alkalis, forming a salt and water:

$$HCl(aq) + NaOH(aq) \longrightarrow NaCl(aq) + H_2O(l)$$
acid + alkali ⟶ salt + water

Redox reactions

Redox reactions are discussed in detail on pages 42–44.

Acids react with many metals in **redox** reactions:

• the metal is oxidised
• hydrogen is reduced.

Acids react with metals, forming a salt and hydrogen gas:

$$2HCl(aq) + Mg(s) \longrightarrow MgCl_2(aq) + H_2(g)$$
$$H_2SO_4(aq) + Mg(s) \longrightarrow MgSO_4(aq) + H_2(g)$$
+1 ────────────────────▶ 0   H reduced
      0 ──────▶ +2   Mg oxidised

acid + metal ⟶ salt + hydrogen

Ammonia as a base

OCR ▶ M1

Ammonia, NH_3 is a colourless gas that is very soluble in water. It forms a weak alkaline solution when dissolved in water.

$$NH_3(aq) + H_2O(l) \rightleftharpoons NH_4^+(aq) + OH^-(aq)$$

Much of the ammonia formed in the Haber process (see pages 113–114) is neutralised by acids to form ammonium salts, such as ammonium nitrate:

$$NH_3(aq) + HNO_3(aq) \longrightarrow NH_4NO_3(aq)$$

Ammonium salts such as ammonium nitrate are very soluble in water and have a high nitrogen content. They are extensively used as artificial fertilisers.

Water of crystallisation

CCEA   M1
OCR   M1

Hydrated crystals

Many salts form crystals, which contain water in their structure. The water can be easily driven off by heat, often forming a non-crystalline powder.

- The crystalline form containing water molecules is **hydrated**.
- The powdered form without water is **anhydrous**.
- The molar proportion of water in the hydrated crystals is **water of crystallisation**.

Notice how water of crystallisation is shown with a '•' in the formula of the hydrated form.

In hydrated copper sulfate:

- the molar proportion of copper sulfate : water is 1 : 5
- the formula is shown as $CuSO_4 \cdot 5H_2O$

Calculating the formula of a hydrated salt

The formula of a hydrated salt can be determined by heating the hydrated salt to drive off the water of crystallisation.

Example

From an experiment to determine the formula of hydrated sodium sulfate,

mass of the **hydrated** salt, $Na_2SO_4 \cdot xH_2O$ = 7.08 g
mass of the **anhydrous** salt, Na_2SO_4 = 3.12 g
∴ mass of H_2O in $Na_2SO_4 \cdot xH_2O$ = 3.96 g

Calculate the amount (in mol) of anhydrous Na_2SO_4:

Molar mass, $M(Na_2SO_4) = 23.0 \times 2 + 32.1 + (16 \times 4) = 142.1$ g mol^{-1}

Remember $n = \dfrac{mass}{Molar\ mass}$

$$n(Na_2SO_4) = \frac{3.12}{142.1} = 0.0220 \text{ mol}$$

Calculate the amount (in mol) of water:

$$n(H_2O) = \frac{3.96}{18.0} = 0.220 \text{ mol}$$

Determine the formula of the hydrated salt:

Molar ratio $Na_2SO_4 : H_2O = 0.0220 : 0.220 = 1 : 10$

∴ formula of the hydrated salt = $Na_2SO_4 \cdot 10H_2O$

Progress check

1 Write equations for the following reactions:
   (a) sulfuric acid and potassium hydroxide
   (b) nitric acid and copper(II) carbonate
   (c) hydrochloric acid and zinc

2 From the experimental results, work out the formula of the hydrated salt.
   Mass of $FeSO_4 \cdot xH_2O$ = 8.34 g
   Mass of $FeSO_4$ = 4.56 g

2 $FeSO_4 \cdot 7H_2O$

1 (a) $H_2SO_4(aq) + 2KOH(aq) \longrightarrow K_2SO_4(aq) + 2H_2O(l)$
  (b) $2HNO_3(aq) + CuCO_3(s) \longrightarrow Cu(NO_3)_2(aq) + H_2O(l) + CO_2(g)$
  (c) $2HCl(aq) + Zn(s) \longrightarrow ZnCl_2(aq) + H_2(g)$

Sample question and model answer

A student carried out an investigation using a sample of hydrochloric acid, HCl(aq).

(a) The concentration of the hydrochloric acid was first determined by titration of a 25.0 cm³ sample against 0.148 mol dm⁻³ sodium hydroxide of which 28.40 cm³ were required.

Calculate the concentration, in mol dm⁻³, of the hydrochloric acid.

no. of moles of NaOH = $\frac{0.148 \times 28.40}{1000}$ = 4.20 × 10⁻³ ✓

NaOH + HCl ⟶ NaCl + H₂O
1 mol NaOH reacts with 1 mol HCl

∴ 4.20 × 10⁻³ mol NaOH reacts with 4.20 × 10⁻³ mol HCl ✓
25.0 cm³ HCl(aq) contains 4.20 × 10⁻³ mol HCl

1 cm³ HCl(aq) contains $\frac{4.20 \times 10^{-3}}{25}$ mol HCl

1 dm³ (1000 cm³) HCl(aq) contains $\frac{4.20 \times 10^{-3}}{25}$ × 1000 = 0.168 mol HCl.

∴ concentration of HCl(aq) is 0.168 mol dm⁻³ ✓

[3]

(b) The student took a 300 cm³ sample of the same hydrochloric acid. The student intended to neutralise the hydrochloric acid with calcium hydroxide, Ca(OH)₂. The equation is shown below:

$$Ca(OH)_2 + 2HCl \longrightarrow CaCl_2 + 2H_2O$$

Calculate the mass, in g, of calcium hydroxide, Ca(OH)₂ required to neutralise this hydrochloric acid.

no. of moles of 0.168 mol dm⁻³ HCl in 300 cm³ = $\frac{0.168 \times 300}{1000}$ = 0.0504 ✓
1 mol Ca(OH)₂ reacts with 2 mol HCl

∴ $\frac{0.0504}{2}$ mol Ca(OH)₂ reacts with 0.0504 mol HCl ✓

Molar mass of Ca(OH)₂ = 40 + (16 + 1)2 = 74.1 g mol⁻¹ ✓

Mass of Ca(OH)₂ = $\frac{0.0504}{2}$ × 74.1 = 1.87 g (to 3 significant figures) ✓
[4]

(c) The student made some calcium hydroxide by heating calcium carbonate and then adding water:
$$CaCO_3 \longrightarrow CaO + CO_2$$
$$CaO + H_2O \longrightarrow Ca(OH)_2$$

Calculate the mass of calcium carbonate that the student would need to produce 10.0 g of calcium hydroxide.

In 10.0 g Ca(OH)₂, number of moles of Ca(OH)₂ = $\frac{10.0}{74.1}$ = 0.135 ✓

From the equations, 1 mol CaCO₃ ⟶ 1 mol CaO ⟶ 1 mol Ca(OH)₂
∴ number of moles of CaCO₃ = 0.135
∴ Mass of CaCO₃ = 0.135 × 100.1 = 13.5 g (to 3 significant figures) ✓
[2]
[Total: 9]

Practice examination questions

1   Chemists use a model of an atom that consists of sub-atomic particles: protons, neutrons and electrons.

(a) Complete the table below to show the properties of these sub-atomic particles.

particle	relative mass	relative charge
proton	1	+1
neutron	1	0
electron	0	-1

[3]

(b) The particles in each pair below differ **only** in the number of protons **or** neutrons **or** electrons. Explain what the difference is within each pair.
   (i) ^{12}C and ^{13}C *neutrons .*
   (ii) ^{16}O and $^{16}O^{2-}$ *electros .*
   (iii) $^{23}Na^+$ and $^{24}Mg^{2+}$.   *electroy , neutrons protons.*   [6]

[Total: 9]

2   (a) In terms of the numbers of sub-atomic particles, state **one** difference *different* and **two** similarities between two isotopes of the same element.   [3]   *N.*
   Same element , same proton
   (b) Give the chemical symbol, including its mass number, for an atom which has 5 electrons and 6 neutrons.   *B .*   [1]

   (c) An element has an atomic number of 15 and it forms an ion with a charge of 3−.
       (i) Deduce the electron configuration of this ion.   *[Ne] $3s^2 3p^3$.*
       (ii) Which block in the Periodic Table is this element in?   *p .*   [2]

   (d) (i) Write an equation for the process involved in the first ionisation energy of silicon.   *$Si + e^- \rightarrow Si$.*
       (ii) Explain why the fifth ionisation energy of silicon is much greater than the fourth.   *fifth requires femainy electron*   [6]
       from +5 ion which is much stonger   [Total: 12]
       than +4 ion.

3   (a) A student added 0.117 g of lithium to 200 cm³ of water. The following reaction took place.

$$2Li(s) + 2H_2O(l) \longrightarrow 2LiOH(aq) + H_2(g)$$

   $\dfrac{0.117}{6.9} = 0.016$ n

   (i) Calculate the number of moles of lithium that reacted.

   (ii) Calculate the concentration, in mol dm⁻³, of the 200 cm³ of aqueous lithium hydroxide solution that formed.

   Pv = nrt .

   (iii) Calculate the volume of hydrogen gas produced at room temperature and pressure (298 K and 100 kPa).   *$V = \dfrac{nrt}{p}$.*   [6]

   (b) In another experiment 25.0 cm³ of 0.126 mol dm⁻³ lithium hydroxide were neutralised by 21.25 cm³ of sulfuric acid as shown below.

$$2LiOH(aq) + H_2SO_4(aq) \longrightarrow Li_2SO_4(aq) + 2H_2O(l)$$

   Calculate the concentration, in mol dm⁻³, of the sulfuric acid.

   [3]
   [Total: 9]

$c = n/v$.

$\left(\dfrac{200 \times 0.016}{1000} \right) =$

$\dfrac{0.016}{(200/1000)} =$

0.2

0.08

$\dfrac{0.0.16}{2} = H_2$ mol.

$\dfrac{8 \times 10^{-3} \times 8.31 \times 298}{100 000} = 1.98 \times 10^{-4} . = 0.19 \, \text{dm}^3 .$

Practice examination questions

4 A student investigated different methods to neutralise 0.580 mol dm^{-3} sulfuric acid, H_2SO_4.

(a) The student decided to add 1.25 mol dm^{-3} NaOH to a 25.0 cm^3 sample of the 0.580 mol dm^{-3} sulfuric acid. The reaction that takes place is shown below.

$$2NaOH(aq) + H_2SO_4(aq) \longrightarrow Na_2SO_4(aq) + 2H_2O(l)$$

    (i) Calculate the amount, in moles, of H_2SO_4 in 25.0 cm^3 of 0.580 mol dm^{-3} of sulfuric acid.

    (ii) What is the amount, in moles, of NaOH that is needed to neutralise this amount of acid?

    (iii) Calculate the volume, in cm^3, of 1.25 mol dm^{-3} NaOH that would be needed to neutralise this amount of acid.

[3]

(b) The student then decided to add sodium carbonate, Na_2CO_3, to a second 25.0 cm^3 sample of the 0.580 mol dm^{-3} sulfuric acid. The equation for this reaction is shown below.

$$Na_2CO_3(s) + H_2SO_4(aq) \longrightarrow Na_2SO_4(s) + H_2O(l) + CO_2(g)$$

Calculate the mass of Na_2CO_3 that would be needed to neutralise this amount of sulfuric acid.

[2]

(c) In a third experiment the student added magnesium metal to the sulfuric acid. The equation for the reaction that takes place is shown below.

$$Mg(s) + H_2SO_4(aq) \longrightarrow MgSO_4(aq) + H_2(g)$$

Use oxidation numbers to show that this is a redox reaction.

[2]

[Total: 7]

5 Sodium azide, NaN_3, decomposes when heated to release nitrogen gas.

$$2NaN_3(s) \longrightarrow 2Na(l) + 3N_2(g)$$

(a) A student prepared 1.50 dm^3 of nitrogen in the laboratory by this method. This gas volume was measured at room temperature and pressure (r.t.p.).

    (i) How many moles of N_2 did the student prepare?
       [Assume that 1 mole of gas molecules occupies 24.0 dm^3 at r.t.p.]

    (ii) What mass of NaN_3 did the student heat? [4]

(b) In this reaction, 0.96 g of sodium metal was also formed. The student carefully reacted the sodium with water to form 50.0 cm^3 of aqueous sodium hydroxide:

$$2Na(s) + 2H_2O(l) \longrightarrow 2NaOH(aq) + H_2(g)$$

    (i) Calculate the concentration, in mol dm^{-3}, of the aqueous sodium hydroxide.

    (ii) Use oxidation numbers to show that this is a redox reaction. [4]

[Total: 8]

Chapter 2

Bonding, structure and the Periodic Table

The following topics are covered in this chapter:

- *Chemical bonding*
- *Shapes of molecules*
- *Electronegativity, polarity and polarisation*
- *Intermolecular forces*
- *Bonding, structure and properties*

- *The modern Periodic Table*
- *The s-block elements: Group 1 and Group 2*
- *The Group 7 elements and their compounds*
- *Extraction of metals*
- *Qualitative analysis*

2.1 Chemical bonding

After studying this section you should be able to:

- *appreciate that compounds often contain atoms or ions with electron structures of a noble gas*
- *understand the nature of an ionic bond in terms of electrostatic attraction between ions*
- *determine the formula of an ionic compound from the ionic charges*
- *understand evidence for ionic bonds from lattice enthalpies*
- *understand the nature of polarisation in ionic compounds*
- *understand the nature of a covalent bond and a dative covalent (coordinate) bond in terms of the sharing of an electron pair*
- *construct 'dot-and-cross' diagrams to demonstrate ionic and covalent bonding*
- *describe metallic bonding*

LEARNING SUMMARY

Why do atoms bond together?

AQA	M1	OCR	M1
EDEXCEL	M1	SALTERS	M1
CCEA	M1	WJEC	M2

The Octet Rule

Under normal conditions, the only elements that can exist as single atoms are the Noble Gases in Group 0 of the Periodic Table. The atoms of all other elements are bonded together. To find out why, we need to look at the electron configurations of the atoms of the Noble Gases.

The Noble Gases get their name from their unreactivity.

In the atoms of a noble gas:

- all electrons are paired,
- the bonding shells are full.

> Atoms of the Noble Gases are stable because all their electrons are paired.

He atom
$1s^2$

Ne atom
$1s^2 2s^2 2p^6$

Ar atom
$1s^2 2s^2 2p^6 3s^2 3p^6$

This arrangement of electrons is particularly stable and is often referred to as a stable 'octet' (as in neon and argon).

> This tendency to acquire a noble gas electron structure is often referred to as the 'Octet Rule'.

Apart from the Noble Gases, atoms are bonded together. Thus the elements oxygen and nitrogen exists as O_2 and N_2. Compounds exist when atoms of different elements bond together as in CO_2, CO and NaCl. In all these examples, electrons have been shared, transferred or rearranged in such a way that the atoms of the elements have acquired a noble gas electron structure. If this creates an outer shell of 8 electrons (as in all these cases), the *'Octet Rule'* is obeyed. All these themes are discussed in the sections that follow.

Types of bonding

Many mistakes are made in chemistry by confusing the type of bonding.

Covalent and ionic compounds are bonded differently and they behave differently. You cannot apply 'ionic bonding ideas' to a compound that is covalent.

Chemical bonds are classified into three main types: ionic, covalent and metallic.

- **Ionic** bonding occurs between the atoms of a **metal** and a **non-metal**. For example, $NaCl$, MgO, Fe_2O_3.
- **Covalent** bonding occurs between the atoms of **non-metals**. For example, O_2, H_2, H_2O, diamond and graphite.
- **Metallic** bonding occurs between the atoms of **metals**. For example, all metals such as iron, zinc, aluminium, etc.; alloys as in brass (copper and zinc), bronze (copper and tin).

> **KEY POINT**
>
> Always decide the type of bonding before:
> - drawing a 'dot-and-cross' diagram
> - predicting how a compound reacts
> - predicting the properties of a compound.

Ionic bonding

AQA	M1	OCR	M1
EDEXCEL	M1	SALTERS	M1
CCEA	M1	WJEC	M2

Ionic bonds

Ionic bonds are present in a compound of a metal and a non-metal.

An ionic bond is formed when electrons are **transferred** from a metal atom to a non-metal atom forming **oppositely-charged ions**.

> **KEY POINT**
>
> An ionic bond is the electrical attraction between oppositely charged ions.

A positive ion is called a **cation** because it is attracted to a **cathode** (a negative electrode). A negative ion is called an **anion** because it is attracted to an **anode** (a positive electrode).

For showing chemical bonding, 'dot-and-cross' diagrams often only show the outer shell.

Example: sodium chloride, NaCl

Sodium chloride, NaCl, is formed by the transfer of one electron from a sodium atom ($[Ne]3s^1$) to a chlorine atom ($[Ne]3s^23p^5$) with the formation of ions:

A sodium atom has lost an electron to acquire the electron structure of neon:

$$Na \longrightarrow Na^+ + e^-$$
$$[Ne]3s^1 \longrightarrow [Ne] + e^-$$

The ions that are formed have stable noble gas electron structures with full outer electron shells.

A chlorine atom has gained an electron to acquire the electron structure of argon:

$$Cl + e^- \longrightarrow Cl^-$$
$$[Ne]3s^23p^5 + e^- \longrightarrow [Ne]3s^23p^6 \text{ or } [Ar]$$

Example: magnesium chloride, MgCl₂

A magnesium atom ($[Ne]3s^2$) has two electrons in its outer shell. One electron is transferred to each of two chlorine atoms ($[Ne]3s^23p^5$) forming ions.

The Mg^{2+} and Cl^- ions have noble gas electron configurations.

Giant ionic lattices

Each ion is surrounded by oppositely-charged ions, forming a giant ionic lattice.

Edexcel studies evidence for the model of ionic bonding in detail. See pages 57–58.

- Although it is convenient to look at ionic bonding between two ions only, each ion is able to attract oppositely charged ions in all directions.
- This results in a **giant ionic lattice** structure comprising hundreds of thousands of ions (depending upon the size of the crystal).
- This arrangement is characteristic of all ionic compounds.

Part of the sodium chloride lattice

- Na$^+$
- Cl$^-$

- Each Na$^+$ ion surrounds 6 Cl$^-$ ions
- Each Cl$^-$ ion surrounds 6 Na$^+$ ions

Ionic charges and the Periodic Table

An ionic charge can be predicted from an element's position in the Periodic Table.

The diagram below shows elements in Periods 2 and 3 of the Periodic Table.

group	1	2	3	4	5	6	7	0
number of outer shell electrons	1	2	3	4	5	6	7	8
element	Li	Be	B	C	N	O	F	Ne
ion	Li$^+$				N^{3-}	O^{2-}	F$^-$	
element	Na	Mg	Al	Si	P	S	Cl	Ar
ion	Na$^+$	Mg^{2+}	Al^{3+}		P^{3-}	S^{2-}	Cl$^-$	

Remember that any metal ions will be positive and non-metal ions negative.

- Metals in Groups 1–3 **lose** sufficient electrons from their atoms to form ions with the electron configuration of the **previous** noble gas;
- Non-metals in Groups 5–7 **gain** sufficient electrons to their atoms to form ions with the electron configuration of the **next** noble gas.
- Be, B, C and Si do not normally form ionic compounds because of the large amount of energy required to transfer electrons forming their ions.

Groups of covalent-bonded atoms can also lose or gain electrons to give ions such as those shown below.

Learn these

1+	1–		2–		3–	
ammonium NH$_4^+$	hydroxide	OH$^-$	carbonate	CO$_3^{2-}$	phosphate	PO$_4^{3-}$
	nitrate	NO$_3^-$	sulfate	SO$_4^{2-}$		
	nitrite	NO$_2^-$	sulfite	SO$_3^{2-}$		
	hydrogen-carbonate	HCO$_3^-$				

Predicting ionic formulae

Although an ionic compound comprises charged ions, its overall charge is zero.

> In an ionic compound,
> total number of positive charges from positive ions = total number of negative charges from negative ions.
>
> **KEY POINT**

Working out an ionic formula from ionic charges

calcium chloride:	ion	charge	aluminium sulfate:	ion	charge
equalise charges:	Ca²⁺	2+	equalise charges:	Al³⁺	3+
				Al³⁺	3+
	Cl⁻	1–			
	Cl⁻	1–		SO₄²⁻	2–
				SO₄²⁻	2–
				SO₄²⁻	2–
total charge must be zero:		0	total charge must be zero:		0
formula:		$CaCl_2$	formula:		$Al_2(SO_4)_3$

Covalent bonding

AQA	M1	OCR	M1
EDEXCEL	M1	SALTERS	M1
CCEA	M1	WJEC	M2

Covalent bonds

A covalent compound comprises molecules: groups of atoms held together by covalent bonds.

> **A molecule is the smallest part of a covalent compound that can take part in a chemical reaction.**
>
> KEY POINT

A covalent bond is formed when electrons are **shared** rather than transferred.

> **A covalent bond is a shared pair of electrons.**
>
> KEY POINT

Covalent bonds are present in a compound of two non-metals.

Example: hydrogen, H₂

The covalent bond in a hydrogen molecule H₂ forms when two hydrogen atoms ($1s^1$) bond together, each contributing 1 electron to the bond:

In forming the covalent bond, each hydrogen atom in the H₂ molecule has:

- control of 2 electrons – one of its own and the second from the other atom
- the electron configuration of the noble gas helium.

A single covalent bond is written as A–B.

When a covalent bond forms, each atom *often* acquires a stable noble gas electron configuration with a full outer bonding shell.

This diagram shows how two atoms are bonded together by a covalent bond.

Unlike an ionic bond, a covalent bond is *directional* and acts solely between the two atoms involved in the bond. Thus hydrogen exists simply as H₂ molecules.

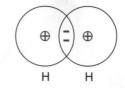

In a hydrogen molecule, the nucleus of each hydrogen atom is attracted towards the electron pair of the covalent bond.

The diagrams below show examples of covalent bonding in some simple molecules.

Notice that each atom contributes one electron to the covalent bond.

Cl₂

H₂O

NH₃

CH₄

Multiple covalent bonds

A covalent bond in which **one** pair of electrons is shared is known as a **single bond**, e.g. Cl_2, written as Cl–Cl.

Atoms can also share more than one pair of electrons to form a **multiple bond**.

- Sharing of **two** pairs of electrons forms a **double bond**, e.g. O_2, written as O=O.
- Sharing of **three** pairs of electrons forms a **triple bond**, e.g. N_2, written as $N \equiv N$.

O_2 N_2

Other common examples include:

C_2H_4: H_2C=CH_2
CO_2: O=C=O
C_2H_2: H–$C \equiv C$–H

Dative covalent or coordinate bonds

A dative covalent bond forms when the shared pair of electrons comes from just **one** of the bonded atoms.

A dative covalent or coordinate bond is one in which **one** of the atoms supplies **both** the shared electrons to the covalent bond. A dative covalent bond is written as A→B, where the direction of the arrow shows the direction in which the electron pair is donated.

Example: ammonium ion, NH_4^+

- The ammonium ion, NH_4^+, forms when ammonia NH_3 bonds with a proton H^+, with both the bonding electrons coming from NH_3.
- Note that the ammonium ion has three covalent bonds and one dative covalent bond:

The involvement of the electron pair as a dative covalent bond is often shown as an arrow:

Note that once NH_4^+ has formed, all 4 bonds are equivalent and you cannot tell which was formed by a dative covalent bond.

NH_3 NH_4^+

In exams, these are easy marks provided that the facts have been learnt.

The Octet Rule doesn't always work

The Octet Rule is useful in chemistry but it can be broken. It is probably more important that all unpaired electrons are paired.

When covalent bonds form, unpaired electrons **often** pair up to form a noble gas electron configuration according to the Octet Rule (see page 51). Many molecules, however, form electron configurations that do not form an octet. Some molecules have 6, 10, 12 or even 14 electrons in the outer bonding shell. These compounds form by the pairing of unpaired electrons.

Phosphorus forms two chlorides, PCl_3 and PCl_5.

PCl_3 obeys the Octet Rule.

In PCl_5,

- the outer shell of phosphorus has expanded allowing all 5 electrons to be used in bonding
- the Octet Rule is broken and phosphorus now has 10 electrons in its outer shell:

In PCl_3, phosphorus is using 3 valence electrons.

In PCl_5, phosphorus is using 5 valence electrons (the same as the Group number of phosphorus).

          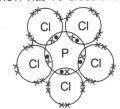

PCl_3: only 3 electrons from the outer bonding shell of phosphorus are used in bonding.

PCl_5: all 5 electrons from the outer bonding shell of phosphorus are used in covalent bonds.

An electron that takes part in forming chemical bonds is called a **valence electron**.

For elements in the s-block and p-block of the Periodic Table, the maximum number of valence electrons available is usually equal to the group number.

Metallic bonding

AQA	M1	OCR	M1
EDEXCEL	M1	SALTERS	M1
CCEA	M1	WJEC	M2

A metallic bond holds atoms together in a solid metal or alloy.

In solid metals, the atoms are ionised.

- The positive ions occupy fixed positions in a lattice.
- The outer shell electrons are **delocalised** – they are spread throughout the metallic structure and are able to move freely throughout the lattice.

> A **metallic bond** is the electrostatic attraction between the positive metal ions and delocalised electrons.

KEY POINT

A metallic lattice is held together by electrostatic attraction between positive metal ions and electrons.

In the metallic lattice, each metal atom exists as + ions by releasing its outer valence electrons to the delocalised pool of electrons (often called a 'sea of electrons').

The 'sea of electrons'

Delocalised and localised electrons

The **delocalised** electrons in metals are spread throughout the metal structure. The delocalisation is such that electrons are able to move throughout the structure and it is impossible to assign any electron to a particular positive ion.

In a covalent bond, the **localised** pair of electrons is always positioned between the two atoms involved in the bond. The electron charge is much more concentrated.

Progress check

1 Draw 'dot-and-cross' diagrams of (a) MgO; (b) Na_2O.

2 Using the information on page 53, predict formulae for the following ionic compounds:
   (a) lithium chloride   (e) sodium carbonate
   (b) potassium sulfide   (f) calcium hydroxide
   (c) lithium nitride   (g) aluminium sulfate
   (d) aluminium oxide   (h) ammonium phosphate.

3 Draw 'dot-and-cross' diagrams for
   (a) C_2H_6;   (b) HCN;   (c) H_3O^+.

2 (a) $LiCl$; (b) K_2S; (c) Li_3N; (d) Al_2O_3; (e) Na_2CO_3; (f) $Ca(OH)_2$; (g) $Al_2(SO_4)_3$; (h) $(NH_4)_3PO_4$.

Testing the model of ionic bonding

EDEXCEL M1

Lattice enthalpy

Lattice enthalpy indicates the strength of the ionic bonds in an ionic lattice.

> **The lattice enthalpy ($\Delta H^{\ominus}_{L.E.}$)** of an ionic compound is the enthalpy change that accompanies the formation of 1 mole of an ionic compound from its constituent gaseous ions. ($\Delta H^{\ominus}_{L.E.}$ is exothermic.)
>
> e.g. $Na^+(g) + Cl^-(g) \longrightarrow Na^+Cl^-(s)$

KEY POINT

Determination of lattice enthalpies

For Hess' Law see page 99.

Lattice enthalpies cannot be determined directly and must be calculated indirectly using Hess' Law from other enthalpy changes that can be found experimentally. The energy cycle used to calculate a lattice enthalpy is the **Born-Haber cycle**.

The basis of the Born-Haber cycle is the formation of an ionic lattice from its elements. For sodium chloride, one route is the single stage process corresponding to its enthalpy change of formation.

ROUTE 1
$$Na(s) + \tfrac{1}{2}Cl_2(g) \longrightarrow Na^+Cl^-(s)$$

A second route can be broken down into separate steps resulting in:

ROUTE 2
* production of gaseous atoms:
$$Na(s) \longrightarrow Na(g)$$
$$\tfrac{1}{2}Cl_2(g) \longrightarrow Cl(g)$$
* production of gaseous ions:
$$Na(g) \longrightarrow Na^+(g) + e^-$$
$$Cl(g) + e^- \longrightarrow Cl^-(g)$$
* production of the solid ionic lattice:
$$Na^+(g) + Cl^-(g) \longrightarrow Na^+Cl^-(s)$$

The Born-Haber cycle

The complete Born-Haber cycle for sodium chloride is shown below. Definitions for these enthalpy changes are shown on page 58.

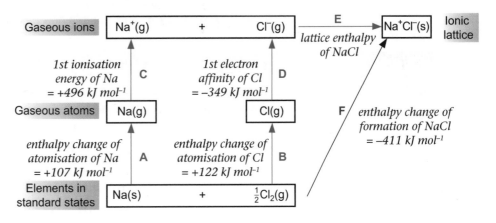

For NaCl, the production of gaseous ions from gaseous atoms comprises two changes:

the 1st ionisation energy of Na is **endothermic**

the 1st electron affinity of Cl is **exothermic**.

Route 1: $A + B + C + D + E$

Route 2: F

Using Hess' Law: $A + B + C + D + E = F$

$\therefore$ $+107 + 122 + 496 + (-349) + E = -411$

Hence, the lattice energy of $Na^+Cl^-(s)$, $E = $ **-787** kJ mol^{-1}

Definitions for enthalpy changes

The enthalpy changes involved in these two routes are shown below.

> **The standard enthalpy change of formation $\Delta H^{\ominus}_f$ is the enthalpy change that takes place when one mole of a compound in its standard state is formed from its constituent elements in their standard states under standard conditions.**
>
> **KEY POINT**

$$Na(s) + \tfrac{1}{2}Cl_2(g) \longrightarrow Na^+Cl^-(s) \qquad \Delta H^{\ominus}_f = -411 \text{ kJ mol}^{-1}$$

> Be careful when using $\Delta H^{\ominus}_{at}$ involving diatomic elements.
>
> $\Delta H^{\ominus}_{at}$ relates to the formation of 1 mole of atoms. For chlorine, this involves $\tfrac{1}{2}Cl_2$ only.

> **The standard enthalpy change of atomisation, $\Delta H^{\ominus}_{at}$ of an element is the enthalpy change that accompanies the formation of 1 mole of gaseous atoms from the element in its standard state.**
>
> **KEY POINT**

$$Na(s) \longrightarrow Na(g) \qquad \Delta H^{\ominus}_{at} = +107 \text{ kJ mol}^{-1}$$
$$\tfrac{1}{2}Cl_2(g) \longrightarrow Cl(g) \qquad \Delta H^{\ominus}_{at} = +122 \text{ kJ mol}^{-1}$$

> **The first ionisation energy $\Delta H^{\ominus}_{I.E.}$ of an element is the enthalpy change that accompanies the removal of 1 electron from each atom in 1 mole of gaseous atoms to form 1 mole of gaseous 1+ ions.**
>
> **KEY POINT**

$$Na(g) \longrightarrow Na^+(g) + e^- \qquad \Delta H^{\ominus}_{I.E.} = +496 \text{ kJ mol}^{-1}$$

> Be careful using electron affinities. $\Delta H^{\ominus}_{E.A.}$ relates to the formation of 1 mole of 1$^-$ ions. For chlorine, this involves $Cl(g)$ only

> **The first electron affinity $\Delta H^{\ominus}_{E.A.}$ of an element is the enthalpy change that accompanies the addition of 1 electron to each atom in 1 mole of gaseous atoms to form 1 mole of gaseous 1$^-$ ions.**
>
> **KEY POINT**

$$Cl(g) + e^- \longrightarrow Cl^-(g) \qquad \Delta H^{\ominus}_{E.A.} = -349 \text{ kJ mol}^{-1}$$

Limitations of lattice enthalpies

The ionic model for ionic bonding can be tested by comparing 'Born-Haber' lattice enthalpies with theoretical values, calculated by considering each ion as a charged sphere. Any difference between a **theoretical** lattice enthalpy and lattice enthalpy calculated using a Born-Haber cycle indicates a proportion of covalent bonding caused by **polarisation**. This is explained below.

Polarisation in ionic compounds

EDEXCEL   M1
CCEA   M2
WJEC   M2

A 100% ionic bond would only form by complete transfer of electrons from a metallic atom to a non-metallic atom. Many ionic compounds have a degree of covalency, caused by polarisation and resulting in an incomplete transfer of electrons.

In Period 3, from Na to Al, the radius of the positive ions decreases.

- The cations Na^+, Mg^{2+} and Al^{3+} are isoelectronic (have the same number of electrons).
- The number of protons increases.

The electrons are therefore attracted closer to the nucleus, increasing the charge density.

> Polarisation is most apparent with a small, densely charged 3$^+$ cation.

Na^+	Mg^{2+}	Al^{3+}
$10e^-$	$10e^-$	$10e^-$
$11p^+$	$12p^+$	$13p^+$

Charge/size ratio increases

Cations and Anions

During electrolysis, the negative electrode or *cathode* attracts the positive ions, called **cations**.

The positive electrode or *anode* attracts the negative ions, called **anions**.

Cations are smaller than their atoms.

Anions are larger than their atoms.

Down a group, the ionic radius increases. For example, the ionic radius of the halide ions, from Cl^- to I^-, increases in size as shells are added, producing less charge density.

Polarisation is greatest between:

- a small + ion (cation) with multiple + charges (i.e. a high charge density) and
- a large – ion (anion) with a low charge density.

The electric field around a **small cation** distorts the electron shells around a **large anion**. As a result, electrons are attracted towards the cation, giving a degree of covalency (electron sharing). This is called **polarisation**.

Small Al^{3+} ion is highly charged.
O^{2-} ion is distorted and polarised resulting in partial covalent character.

Stability of ionic compounds

EDEXCEL   M1

The stability of ionic compounds can be compared using values calculated for enthalpy changes of formation based on Born-Haber cycles.

e.g. $MgCl$: -94 kJ mol^{-1}; $MgCl_2$ -643 kJ mol^{-1}; $MgCl_3$ $+3943$ kJ mol^{-1}

$MgCl_2$ is the compound that actually forms: it is the most stable magnesium chloride because it has the most exothermic enthalpy change of formation.

Progress check

1 Using the information below:

(a) name each enthalpy change **A** to **E**

(b) construct a Born-Haber cycle for sodium bromide and calculate the lattice enthalpy of sodium bromide. (Use the Born-Haber cycle for NaCl on page 57 as a guide.)

enthalpy change	equation	$\Delta H^\ominus$ / kJ mol^{-1}
A	$Na(s) + \frac{1}{2}Br_2(l) \longrightarrow NaBr(s)$	-361
B	$Br(g) + e^- \longrightarrow Br^-(g)$	-325
C	$Na(s) \longrightarrow Na(g)$	$+107$
D	$Na(g) \longrightarrow Na^+(g) + e^-$	$+496$
E	$\frac{1}{2}Br_2(l) \longrightarrow Br(g)$	$+112$

1 (a) A enthalpy change of formation of sodium bromide
        B 1st electron affinity of bromine
        C enthalpy change of atomisation of sodium
        D 1st ionisation energy of sodium
        E enthalpy change of atomisation of bromine
   (b) -751 kJ mol^{-1}

2.2 Shapes of molecules

After studying this section you should be able to:

- *use electron-pair repulsion to explain the shapes and bond angles of simple molecules*
- *understand that lone pairs have a larger repulsive effect than bonded pairs, leading to distortion of the molecular shape*
- *understand how multiple bonds affect the shape of a molecule*

Electron-pair repulsion theory

AQA	M1	OCR	M1
EDEXCEL	M2	SALTERS	M1
CCEA	M1	WJEC	M2

Electron-pair repulsion theory states that:

- the shape of a molecule depends upon the number of electron pairs surrounding the central atom
- electron pairs repel one another and move as far apart as possible.

The shape of any molecule can be predicted by applying this theory.

> Bonds and lone pairs are negative charge clouds that repel one another.

Molecules with bonded pairs

> Learn these shapes and bond-angles.

You can predict the shape of a molecule from its 'dot-and-cross' diagram. Try to relate the number of electron pairs in the examples below with the three-dimensional shape of each molecule.

molecule	$BeCl_2$	BF_3	CH_4	PCl_5	SF_6
dot-and-cross diagram					
electron pairs around central atom	2	3	4	5	6
shape and bond angle					
name of shape	linear	trigonal planar	tetrahedral	trigonal bipyramid	octahedral

Molecules with lone pairs

Each of the examples on page 61 are molecules with four electron pairs surrounding the central atom. The molecular shape will therefore be based upon a tetrahedron. However, a lone pair is closer to an atom than a bonded pair of electrons and has a larger repulsive effect than a bonded pair.

The relative magnitudes of electron-pair repulsions are:

lone-pair/lone-pair > bonded-pair/lone-pair > bonded-pair/bonded-pair.

Lone pairs distort the shape of a molecule and reduce the bond angle. Look at the examples below and notice how each lone pair decreases the bond-angle by 2.5°.

Always draw 3-D shapes – you may be penalised in exams otherwise.

molecule	CH_4	NH_3	H_2O
'dot-and-cross' diagram			
number of lone pairs	0	1	2
shape and bond angle	109.5°	107°	104.5°
name of shape	tetrahedral	pyramidal	non-linear

Molecules with double bonds

In molecules containing multiple bonds, each double bond is treated in the same way as a bonded pair. In the diagram of carbon dioxide below, each double bond is treated as a *'bonding region'*.

molecule	*'dot-and-cross' diagram*	*number of bonding regions*	*shape and bond angle*	*name of shape*
CO_2		2	180° $O{=}C{=}O$	linear

Progress check

1 For each of the following molecules, predict the shape and bond angles.
- (a) BeF_2
- (b) $AlCl_3$
- (c) SiH_4
- (d) H_2S
- (e) PH_3
- (f) CS_2
- (g) SO_3
- (h) SO_2

1 (a) linear, 180°
(b) trigonal planar, 120°
(c) tetrahedral, 109.5°
(d) non-linear, 104.5°
(e) pyramidal, 107°
(f) linear, 180°
(g) trigonal planar, 120°
(h) non-linear, 120°.

2.3 Electronegativity, polarity and polarisation

After studying this section you should be able to:

- *appreciate that many chemical bonds have bonding intermediate between ionic and covalent bonding*
- *describe electronegativity in terms of attraction for bonding electrons*
- *understand the nature of polarity in molecules of covalent compounds*

Ionic or covalent?

AQA	M1	OCR	M1
EDEXCEL	M2	SALTERS	M2
CCEA	M1	WJEC	M2

Between the extremes of ionic and covalent bonding, there is a whole range of intermediate bonds, which have both ionic and covalent contributions.

An ionic bond with 100% ionic character would require the complete transfer of an electron from a metal atom to a non-metal atom. In practice, this never completely happens, although some compounds have close to 100% ionic character.

In a hydrogen molecule, H_2, the two hydrogen atoms are identical and form a covalent bond with 100% covalent character by equally sharing the bonded pair of electrons. However, in molecules such as HCl, the bonded pair of electrons is not shared equally.

An ionic bond is often formed with some covalent character:
- the electron transfer is incomplete
- there is a degree of electron sharing.

A covalent bond is often formed with some ionic character:
- the electrons are not equally shared
- there is a degree of electron transfer.

The sections that follow discuss intermediate bonding in covalent compounds.

Electronegativity

AQA	M1	OCR	M1
EDEXCEL	M2	SALTERS	M2
CCEA	M1	WJEC	M2

The nuclei of the atoms in a molecule attract the electron pair in a covalent bond.

> **Electronegativity is a measure of the attraction of an atom in a molecule for the pair of electrons in a covalent bond.**
>

The most electronegative atoms attract bonding electrons most strongly.

- In general, small atoms are electronegative atoms.
- The most electronegative atoms are those of highly reactive non-metallic elements (such as O, F and Cl).
- Reactive metals (such as Na and K) have the least electronegative atoms.

How is electronegativity measured?

Notice that the Noble Gases are not included. Although neon and helium atoms are smaller than those of fluorine, they do not form bonds and so they have no affinity for bonded electrons.

The 'Pauling scale' is often used to compare the electronegativities of different elements. The diagram below shows how the electronegativity of an element relates to its position in the Periodic Table. The numbers give the Pauling electronegativity values.

See how the Pauling values of electronegativity relate to the element's position in the Periodic Table.

electronegativity increases						
Li	Be	B	C	N	O	F
1.0	1.5	2.0	2.5	3.0	3.5	4.0
Na						Cl
0.9						3.0
K						Br
0.8						2.8

Fluorine, the most electronegative element, has small atoms with a greater attraction for the pair of electrons in a covalent bond than larger atoms.

The greater the **difference** between electronegativities, the greater the **ionic** character of the bond.

The greater the **similarity** in electronegativities, the greater the **covalent** character of the bond.

Polar and non-polar molecules

AQA	M1	OCR	M1
EDEXCEL	M2	SALTERS	M2
CCEA	M1	WJEC	M2

When bonding atoms are the same, the attraction for the bonded pair of electrons is the same and the bond is non-polar.

Hydrocarbons are non-polar because C and H have very similar electronegativities.

In a polar covalent bond, the bonded electrons are attracted towards the more electronegative of the atoms.

The electronegativities of hydrogen halides are discussed in more detail on page 80.

Dipoles in symmetrical molecules cancel.

Non-polar bonds

A covalent bond is non-polar when:

- the bonded electrons are shared **equally** between both atoms
- the bonded atoms have similar electronegativities.

A covalent bond **must** be non-polar if the bonded atoms are the same, as in molecules of H_2 and Cl_2 shown here:

H_2 molecule

Cl_2 molecule

Polar bonds

A covalent bond is polar when:

- the electrons in the bond are shared **unequally** making a *polar bond*
- the bonded atoms are different, each has a different electronegativity.

In hydrogen chloride, HCl, the Cl atom is more **electronegative** than the H atom, making the H–Cl covalent bond polar.

The molecule is *polarised* with a small positive charge δ+ on the hydrogen atom and a small negative charge δ– on the chlorine atom.

bonded pair of electrons attracted closer to chlorine

The hydrogen chloride molecule is *polar* with a *permanent dipole*.

Symmetrical and unsymmetrical molecules

If the molecule is symmetrical, any dipoles will cancel and the molecule will not have a permanent dipole. The diagram below shows why the symmetrical molecule, CCl_4, is non-polar.

- Each C–Cl bond is polar but the dipoles act in different directions.
- The overall effect is for the dipoles to cancel each other.
- ∴ CCl_4 is a non-polar molecule.

CCl_4 is non-polar

Although each C–Cl bond is polar, the dipoles cancel

Progress check

1 Decide whether each molecule is polar and state which atom, if any, has the δ– charge.
  (a) Br_2; (b) H_2O; (c) O_2; (d) HBr; (e) NH_3.

2 (a) Predict the shape of a molecule of BF_3 and of PF_3.
  (b) Explain why BF_3 is non-polar whereas PF_3 is polar.

2 (a) BF_3: trigonal planar; PF_3: pyramidal.
  (b) BF_3 has a symmetrical shape. Although each bond is polar, the dipoles cancel. PF_3 is not symmetrical. The F atoms are all on the same side of the molecule resulting in a permanent dipole.

1 (a) non-polar; (b) polar, Oδ–; (c) non-polar; (d) polar, Brδ–; (e) polar, Nδ–.

2.4 Intermolecular forces

After studying this section you should be able to:

- *understand the nature of van der Waals' forces (induced dipole-dipole interactions)*
- *understand the nature of permanent dipole-dipole interactions and hydrogen bonds*
- *describe the anomalous properties of water arising from hydrogen bonding.*

Van der Waals' forces or induced dipole-dipole forces

AQA	M1	OCR	M1
EDEXCEL	M2	SALTERS	M2
CCEA	M1	WJEC	M2

Van der Waals' forces (induced dipole-dipole forces) exist between all molecules whether polar or non-polar.

> Induced dipole-dipole interactions are often referred to as van der Waals' forces.

What causes van der Waals' forces?

Van der Waals' forces are:

- weak intermolecular forces
- caused by attractions between very small dipoles in molecules.

The diagram below shows how these forces arise between single atoms in a noble gas.

> Ionic and covalent bonds are of comparable strength.
>
> Intermolecular forces are far weaker.

Movement of electrons produces an oscillating dipole

dipole oscillates and continually changes with time

> An oscillating or instantaneous dipole is caused by an uneven distribution of electrons at an instant of time.

Oscillating dipole induces a dipole in a neighbouring molecule which is induced onto further molecules

induced dipoles attract one another

What affects the strength of van der Waals' forces?

Van der Waals' forces result from interactions of electrons between molecules.

The greater the number of electrons in each molecule:

- the larger the oscillating and induced dipoles
- the greater the attractive forces between molecules
- the greater the van der Waals' forces.

> Van der Waals' forces increase in strength with increasing number of electrons.

This can be seen by comparing the boiling points of the Noble Gases.

Noble gas	b. pt./°C	number of electrons	trend
He	−269	2	
Ne	−246	10	• easier to distort electron clouds
Ar	−186	18	
Kr	−152	36	• induced dipoles increase
Xe	−107	54	
Rn	− 62	86	• boiling point increases

Permanent dipole-dipole interactions

AQA	M1	OCR	M1
EDEXCEL	M2	SALTERS	M2
CCEA	M1	WJEC	M2

Permanent dipole-dipole interactions are 'non-directional'.

They are simply weak attractions between dipole charges on different molecules.

The small δ+ and δ- charges on a polar molecule attract oppositely charged dipoles on another polar molecule.

This gives a weak *intermolecular force* called a permanent dipole-dipole interaction.

Example: Intermolecular forces between HCl molecules

Between hydrogen chloride molecules, there will be both:

• van der Waals' forces and
• permanent dipole-dipole interactions.

Although both are weak forces, the permanent dipole-dipole interactions are still stronger than the van der Waals' forces.

Hydrogen bonds

AQA	M1	OCR	M1
EDEXCEL	M2	SALTERS	M2
CCEA	M1	WJEC	M2

A hydrogen bond is a special type of permanent dipole-dipole interaction found between molecules containing the following groups:

A hydrogen bond is a comparatively strong intermolecular attraction between:

• an electron deficient hydrogen atom, $H^{\delta+}$, on one molecule and
• a lone pair of electrons on a highly electronegative atom of F, O or N on another molecule.

Hydrogen bonding occurs between molecules such as H_2O:

Note the role of the lone pair – this is essential in hydrogen bonding.

A hydrogen bond is shown between molecules as a dashed line.

hydrogen bond formed by attraction between dipole charges on different molecules

Hydrogen bonding is especially important in organic compounds containing –OH or –NH bonds: e.g. alcohols, carboxylic acids, amines, amino acids.

Special properties of water arising from hydrogen bonding

A hydrogen bond has only one-tenth the strength of a covalent bond. However, hydrogen bonding is strong enough to have significant effects on physical properties, resulting in some unexpected properties for water.

The solid (ice) is less dense than the liquid (water)

Solids are usually denser than liquids – but ice is less dense than water.

• Particles in solids are usually packed closer together than in liquids.
• Hydrogen bonds hold water molecules apart in an open lattice structure.

Therefore ice is less dense than water.

When water changes state, the covalent bonds between the H and O atoms in an H_2O molecule are strong and do **not** break.

It is the intermolecular forces that break.

The diagram below shows how the open lattice of ice collapses on melting.

hydrogen bonds break

MELTING OF ICE

tetrahedral open lattice in ice

ice lattice collapses: molecules move closer together

Ice has a relatively high melting point, and water a relatively high boiling point

- There are relatively strong hydrogen bonds between H_2O molecules.
- The hydrogen bonds are extra forces over and above van der Waals' forces.
- These extra forces result in higher melting and boiling points than would be expected from just van der Waals' forces.
- When the ice lattice breaks, hydrogen bonds are broken.

Other properties

The extra intermolecular bonding from hydrogen bonds also explains the relatively high surface tension and viscosity in water.

Evidence for hydrogen bonds

The boiling points of the Group 6 and Group 7 hydrides are shown below.

Group 6		Group 7	
hydride	*boiling point / K*	*hydride*	*boiling point / °C*
H_2O	373	HF	293
H_2S	212	HCl	188
H_2Se	232	HBr	206
H_2Te	271	HI	238

The boiling points increase from $H_2S \longrightarrow H_2Te$ and from $HCl \longrightarrow HI$:

- the number of electrons in molecules increase
- van der Waals' forces increase
- more energy is required to break the intermolecular bonds to vaporise the hydrides.

Without hydrogen bonding, H_2O would be a gas at room temperature and pressure.

The boiling point of the first hydride in each group is higher than expected. This provides evidence that there are some extra forces acting between the molecules that must be broken to boil each hydride. These extra forces are hydrogen bonds.

Progress check

1 Which of the following molecules have hydrogen bonding: H_2O, H_2S, CH_4, CH_3OH, NO_2?

2 Draw diagrams showing hydrogen bonding between:
   (a) 2 molecules of ammonia
   (b) 1 molecule of water and 1 molecule of ethanol.

1 H_2O, CH_3OH.

2 (a)

(b)

2.5 Bonding, structure and properties

After studying this section you should be able to:

- describe the typical properties of an ionic compound in terms of its structure
- describe the typical properties of a covalent compound in terms of simple molecular and giant molecular structures
- describe the structure and associated properties of diamond and graphite
- understand the properties of metals in terms of metallic bonding

LEARNING SUMMARY

Bonds and forces

AQA	M1	OCR	M1
EDEXCEL	M1	SALTERS	M1, M2
CCEA	M1	WJEC	M2

Bonding, structure and properties are all related.

A structure is held together by bonds and forces. The different types of bonds introduced in the last section are shown below.

These are very important and are needed to understand the links between bonding, structure and properties.

> Covalent bonds act between atoms.
> Ionic bonds act between ions.
> Metallic bonds act between positive ions and electrons.
> Hydrogen bonds act between polar molecules.
> Dipole-dipole interactions act between polar molecules.
> Van der Waals' forces act between molecules.

KEY POINT

Ionic, covalent and metallic bonds are of comparable strength.

Intermolecular forces are much weaker, and their relative strengths are compared with the strength of covalent bonds in the table below.

type of bond	bond enthalpy / kJ mol^{-1}
covalent bond	200–500
hydrogen bond	5–40
van der Waals' forces	~2

The properties of a substance depend upon its **bonding** and **structure**.

Properties of ionic compounds

AQA	M1	OCR	M1
EDEXCEL	M1	SALTERS	M1, M2
CCEA	M1	WJEC	M2

Ionic compounds form giant ionic lattices with each ion surrounded by ions of the opposite charge. The ions are held together by **strong** electrostatic attraction between positive and negative ions. (See page 52.)

Ionic compounds are solids at room temperature.

The strong forces between ions result in high melting and boiling points.

High melting point and boiling point
- High temperatures are needed to break the strong electrostatic forces holding the ions rigidly in the solid lattice.

Therefore ionic compounds have high melting and boiling points.

Ionic compounds conduct only when ions are free to move – when molten or in aqueous solution.

Electrical conductivity

In the solid lattice,
- the ions are in a fixed position and there are no mobile charge carriers.

Therefore an ionic compound is a **non-conductor** of electricity in the solid state.

When melted or dissolved in water,
- the solid lattice breaks down
- the ions are now free to move as mobile charge carriers.

Therefore an ionic compound is a **conductor** of electricity in liquid and aqueous states.

Solubility

- The ionic lattice often dissolves in **polar** solvents (e.g. water).
- Polar water molecules break down the lattice and surround each ion in solution as shown below for sodium chloride.

The solubility of ionic compounds in water increases with increasing temperature.

- Water molecules attract Na⁺ and Cl⁻ ions.
- Lattice breaks down as it dissolves.
- Water molecules surround ions.

Na⁺ attracts δ– partial charges on the O atoms of water molecules.

Cl⁻ attracts δ+ partial charges on the H atoms of water molecules.

Solubility curves

WJEC   M2

The solubility of a solute in a solvent typically depends on temperature.

- For most solid solutes, the solubility increases with temperature.

A solubility curve shows how the solubility of a substance varies with temperature.

A solution is saturated when no more of the solute salt dissolves in a solvent at that particular temperature.

Solubility is usually measured as the mass of solute that saturates 100 grams of water at a particular temperature.

The diagram below shows solubility curves of NaCl and KNO_3.

Soluble salts can be recovered from an aqueous solution by crystallisation.

- A solution is saturated with the salt by evaporation of the solvent using heat.
- On cooling, the water can no longer dissolve all of the salt at the lower temperature and the excess salt crystallises out.

For the two salts above, far more KNO_3 can be recovered on cooling because the difference in solubility with temperature is much greater than for NaCl.

Properties of covalent compounds

AQA	M1	OCR	M1
EDEXCEL	M2	SALTERS	M1, M2
CCEA	M1	WJEC	M2

Elements and compounds with covalent bonds have either of two structures:
- a simple molecular structure
- a giant covalent structure.

Simple molecular structures, e.g. iodine, I_2

Simple molecular structures form solid lattices with small molecules, such as Ne, H_2, O_2, N_2, held together by **weak** intermolecular forces.

> Intermolecular forces are weak.
>
> Only the intermolecular forces break when a simple molecule melts or boils.
>
> A common mistake in exams is to confuse intermolecular forces with covalent bonds.

Low melting point and boiling point
- Low temperatures provide sufficient energy to break the weak intermolecular forces between molecules.

Therefore simple molecular structures have low melting and boiling points.

The simple molecular structure of solid I_2

> Giant structure:
> high melting point.
>
> Simple molecular structure:
> low melting point.

- When the I_2 lattice breaks down, only the weak van der Waals' forces between the I_2 molecules break.
- In the I_2 molecule, the covalent bond, I–I, is strong and does **not** break when the lattice breaks down.

Strong covalent bonds hold together each I_2 molecule

Weak van der Waals' forces between I_2 molecules

Electrical conductivity
- There are no mobile charged particles.

Therefore simple molecular structures are **non-conductors** of electricity.

Solubility
- Van der Waals' forces form between a simple molecular structure and a non-polar solvent, such as hexane. These weaken the structure.

Therefore simple molecular structures are often soluble in **non-polar** solvents (e.g. hexane).

Giant covalent structures, e.g. carbon (diamond and graphite)

Giant covalent structures have thousands of atoms bonded together by **strong** covalent bonds. This type of structure is known by a variety of names: a giant **covalent** lattice, a giant **molecular** lattice or a giant **atomic** lattice.

> Diamond, graphite and SiO_2 are examples of giant covalent lattices.

High melting point and boiling point
- High temperatures are needed to break the strong covalent bonds in the lattice.

Therefore giant covalent structures have high melting and boiling points.

> Covalent bonds are strong.
>
> The covalent bonds break when a giant molecular structure melts or boils – this only happens at high temperatures.

Electrical conductivity
- Except for graphite (see page 70), there are no mobile charged particles.

Therefore giant covalent structures are **non-conductors** of electricity.

> Carbon can also form other structures: fullerenes and nanotubes.

Solubility
- The strong covalent bonds in the lattice are too strong to be broken by either polar **or** non-polar solvents.

Therefore giant covalent structures are insoluble in polar *and* non-polar solvents.

> Silicon (IV) oxide, SiO_2, has a similar tetrahedral structure to diamond.

Comparison between the properties of diamond and graphite

property	diamond		graphite	
structure		symmetrical structure held together by strong covalent bonds throughout lattice tetrahedral		strong layer structure but with weak bonds between the layers hexagonal layers
electrical conductivity	*poor* • There are no delocalised electrons as all outer shell electrons are used for covalent bonds.		*good* • Delocalised electrons between layers. • Electrons are free to move parallel to the layers when a voltage is applied.	
hardness	*hard* • Tetrahedral shape enables external forces to be spread out throughout the lattice.		*soft* • Bonding within each layer is strong but weak forces between layers easily allow layers to slide.	

Properties of metals

AQA	M1	OCR	M1
EDEXCEL	M1	SALTERS	M1, M2
CCEA	M1	WJEC	M2

Metals have a giant metallic lattice structures held together by **strong** electrostatic attractions between positive ions and negative electrons.

High melting point and boiling point

• Generally high temperatures are needed to separate the ions from their rigid positions within the lattice.

Therefore most metals have high melting and boiling points.

Good thermal and electrical conductivity

> When a metal conducts electricity, only the electrons move.

• The existence of mobile, delocalised electrons allows metals to conduct heat and electricity well, even in the solid state.
• The electrons are free to flow between positive ions.
• The positive ions do **not** move.

drift of delocalised electrons across potential difference

Modern materials

EDEXCEL	M2
WJEC	M2

Modern research is investigating new types of materials for a whole range of uses. These include nanomaterials and smart materials.

Nanomaterials

Nanomaterials have:

> 1 nanometre,
> 1 nm = 10^{-9} m.

• dimensions on a molecular scale: 1 – 100 nm
• unique properties arising from their nanoscale.

A single-walled carbon nanotube is a single layer of graphite rolled up into a cylinder with a diameter of approximately 1 nm (10^{-9} m). Nanotubes have the potential to carry drugs into cells but extensive uses of nanomaterials have yet to be realised.

Smart materials (WJEC only)

A smart material changes properties with a change in conditions (e.g. temperature, moisture, pH, or electric and magnetic fields) and this is often caused by a change in structure.

Summary of properties from structure and bonding

structure	bonding	melting pt / boiling pt	reason	electrical conductivity	reason	solubility	reason
giant ionic	ionic bonds throughout structure	high	strong electrostatic attraction between oppositely charged ions	poor when solid good when aqueous or molten	ions in a fixed position in lattice lattice has broken down: mobile ions	good in polar solvents, e.g. water	attraction between ionic lattice and polar solvent
simple molecular	covalent bonds **within** molecules, van der Waals' forces **between** molecules	low	weak van der Waals' forces between molecules	poor	no mobile charged particles (electrons or ions)	good in non-polar solvents	van der Waals' forces between molecular structure and solvent
giant covalent	covalent bonds throughout structure	high	strong covalent bonds between atoms	poor	no mobile charged particles (electrons or ions)	poor	forces within lattice too strong to be broken by solvents
hydrogen bonded	hydrogen bonds between molecules	low *but* higher than expected	dipole-dipole attractions between molecules	poor	no mobile charged particles (electrons or ions)	good in polar solvents, e.g. water	attraction between dipoles
giant metallic	metallic bonds throughout structure	usually high	strong electrostatic attractions between ions and electrons	good	mobile electrons, even in solid state	poor	forces within lattice too strong to be broken by solvents

> A high melting point is the result of any giant structure, bonded together with strong forces.
>
> Giant structures can be ionic, covalent or metallic.

KEY POINT

Progress check

1 For (a) MgO; (b) CH_4; (c) SiO_2, state the structure and explain the following physical properties: melting and boiling points, electrical conductivity, solubility.

1 (a) giant ionic: high m.pt/b.pt. – strong forces between ions.
Non-conductor when solid – ions fixed in lattice. Conductor when molten or dissolved in water – ions are able to move.
Soluble in water – dipole on water attracted to ions.
(b) simple molecular: low m. pt/b. pt. – weak van der Waals' forces between molecules.
Non-conductor – no mobile charge carriers, electrons localised in covalent bonds.
Soluble in non-polar solvents – van der Waals' forces in solvent interact with van der Waals' forces in CH_4.
(c) giant molecular: high m. pt/b. pt. – strong covalent bonds between atoms.
Non-conductor – no mobile charge carriers, electrons localised in covalent bonds.
Insoluble in all solvents – strong covalent bonds are too strong to be broken by either polar or non-polar solvents.

2.6 The modern Periodic Table

After studying this section you should be able to:

- describe the Periodic Table in terms of atomic number, periods and groups
- classify the Periodic Table into s, p, d and f blocks
- describe and explain periodic trends in atomic radii, ionisation energies, melting and boiling points, and electrical conductivities

Arranging the elements

AQA	M1	OCR	M1
EDEXCEL	M1	SALTERS	M1
CCEA	M1	WJEC	M2

In the Periodic Table:

- Elements are arranged in order of increasing atomic number
- Groups are vertical columns including elements with similar properties
- Periods are horizontal rows across which there is a trend in properties.

Key areas in the Periodic Table

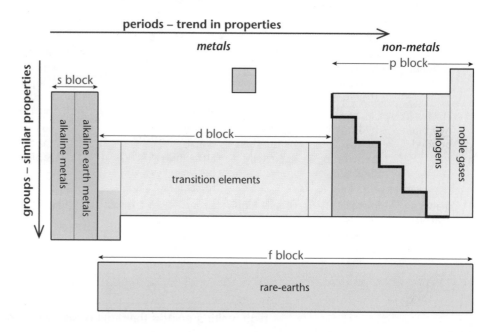

- The diagonal stepped line separates metals (to the left) from non-metals (to the right).
- Elements close to this line, such as silicon and germanium, are called **semi-metals** or **metalloids**. They show properties intermediate between those of a metal and a non-metal.
- The four blocks (s, p ,d and f) show the sub-shell being filled.

Periodicity

> The trend in properties across a period is repeated across each period – this is called *periodicity*.

Periodicity is the periodic trend in properties, repeated across each period.

e.g.   Period 2   METAL   ⟶   NON-METAL
            Period 3   METAL   ⟶   NON-METAL

This periodicity of properties means that predictions can be made about the likely properties of an element and its compounds.

Group 4
C
Si
Ge
Sn
Pb
non-metal to metal

Note, however, the metal/non-metal divide. On descending the Periodic Table, the changeover from metal to non-metal takes place further to the right. For example at the top of Group 4 carbon is a non-metal whereas, at the bottom of Group 4, tin and lead are metals. This means that subtle trends in properties take place down a group.

Trends in atomic radii and first ionisation energies

AQA	M1	OCR	M1
EDEXCEL	M1	SALTERS	M2
CCEA	M1	WJEC	M2

Ionisation energies measure the ease with which an atom of an element loses electrons (see page 26).

Across a period

Across a period, the atomic radius decreases. The trend in atomic radius across Period 2 is shown below:

Li	Be	B	C	N	O	F
$3p^+$	$4p^+$	$5p^+$	$6p^+$	$7p^+$	$8p^+$	$9p^+$

number of protons increases

radius decreases

Atomic radius decreases across a period.

First ionisation energy increases across a period.

Across a period:

- the **nuclear charge increases**
- outer electrons are being added to the same shell
- the attraction between the nucleus and outer electrons increases
- the atomic radius **decreases**
- the first ionisation energy **increases**.

Down a group *Task 4*

Atomic radius increases down a group.

First ionisation energy decreases down a group.

Down a group:

- extra shells are added that are **further** from the nucleus
- there are more shells between the outer electrons and the nucleus leading to **greater shielding** of the nuclear charge
- the attraction between the nucleus and the outer electrons decreases
- the atomic radius **increases**
- the first ionisation energy **decreases**.

shells increase

shielding increases

atomic radius increases

Down a group, the nuclear charge also increases but this is more than compensated by the increase in atomic radius and shielding.

> **KEY POINT**
>
> Across a Period, nuclear charge is the key factor.
>
> Down a Group, atomic radius and shielding are the key factors (see also ionisation energies, page 26).

Trends in melting and boiling points

AQA	M1	OCR	M1
EDEXCEL	M1	SALTERS	M1
CCEA	M1	WJEC	M2

Trends in melting and boiling point provide information about structure: giant or simple molecular.

The graphs below show the variation in boiling points across Period 2 and Period 3 of the Periodic Table.

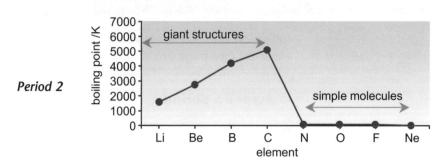

Period 2

Period 3

Across a period:

- the boiling point increases from Group 1 to Group 4
- there is a sharp decrease in boiling point between Group 4 and Group 5
- the boiling points are comparatively low from Group 5 to Group 8.

> **The sharp decrease in boiling point marks a change from giant structures to simple molecular structures.**
>
> **KEY POINT**

Note the periodicity in boiling point: the trend for Period 2 is repeated across Period 3. The trend in melting points is similar.

More details of the structures of these elements are shown below.

Notice the changes in the molecular formulae of P_4, S_8, Cl_2 and Ar. This shows up well in the graph of boiling points across Period 3.

With more atoms (and consequently electrons) in each molecule, the forces between molecules (van der Waals' forces) increase.

When a giant structure is melted or boiled:

- strong forces are broken
- a large input of energy is required
- the melting and boiling points are high.

When a simple molecular structure is melted or boiled:

- weak forces between molecules are broken
- a relatively small input of energy is required
- the melting and boiling points are low.

Comparison of the boiling points of the metals

Note the increase in boiling points of the metals Na → Al in Period 3.

- The number of delocalised electrons in the lattice increases.
- The charge on each cation increases.
- This results in increasing attractive forces within the metallic lattice.

The diagram below compares the attractive forces between atoms of sodium, magnesium and aluminium.

> See also Metallic bonding, pages 56, 70.

ionic charges increases
number of outer shell electrons increases

attraction increases: boiling point increases

The increasing number of delocalised electrons in the lattice also explains the increase in electrical conductivity from Na → Al.

Progress check

1 Using ideas about nuclear charge, attraction and shells, explain the trend in atomic radii across a period and down each group.
2 Why is the melting point of carbon much higher than that of nitrogen?

1 Period: the nuclear charge increases as electrons are being added to the same shell. The attraction between the nucleus and outer electrons increases ∴ the atomic radius decreases across a period. Group: extra shells are added that are further from the nucleus leading to an increased shielding of the outer electrons from the nucleus. Both these factors lead to less attraction between the nucleus and the outer electrons ∴ the atomic radius increases down a group.
2 Carbon has a giant molecular structure; between the atoms there are strong covalent bonds which break on melting. Nitrogen has a simple molecular structure; between the molecules there are weak van der Waals' forces that break on melting.

2.7 The s-block elements: Group 1 and Group 2

After studying this section you should be able to:

- *describe the characteristic properties of the s-block elements*
- *understand how flame tests can be used to identify s-block cations*
- *describe the reactions of the s-block elements with oxygen and with water*
- *recall some of the reactions of s-block oxides*
- *recall the relative solubilities of the Group 2 hydroxides and sulfates*
- *describe and explain the thermal stability of the s-block carbonates and nitrates.*

LEARNING SUMMARY

General properties

AQA	M2	OCR	M1
EDEXCEL	M2	SALTERS	M1, M2
CCEA	M2	WJEC	M2

The elements in Group 1 and Group 2 have hydroxides that are alkaline and their common names reflect this.

- The Group 1 elements are the **Alkali Metals**.
- The Group 2 elements are the **Alkaline Earth Metals**.

Electron configuration

The elements in Group 1 and Group 2 have their highest energy electrons in an s sub-shell and these two groups are known as the *s-block elements*.

Each Group 1 element has:
- **one electron more than the electron configuration of a noble gas**
- **an outer s sub-shell containing one electron.**

Each Group 1 element reacts in a similar way as each atom has 1 electron in the outer shell.

Electron configuration of the s-block elements

Group 1		Group 2	
Li	[He] 2s^1	Be	[He] 2s^2
Na	[Ne] 3s^1	Mg	[Ne] 3s^2
K	[Ar] 4s^1	Ca	[Ar] 4s^2
Rb	[Kr] 5s^1	Sr	[Kr] 5s^2
Cs	[Xe] 6s^1	Ba	[Xe] 6s^2
Fr	[Rn] 7s^1	Ra	[Rn] 7s^2

The s-block elements
Group 1
Group 2

Each Group 2 element has:
- **two electrons more than the electron configuration of a noble gas**
- **an outer s sub-shell containing two electrons.**

Each Group 2 element reacts in a similar way as each atom has 2 electrons in the outer shell.

Physical properties

Group 1
- They are soft metals and can be cut with a knife.
- They have low melting and boiling points.
- They have low densities: Li, Na and K all float on water.
- They have colourless compounds.

Densities of s-block elements

Group 1 Density/g cm^{-3}		Group 2 Density/g cm^{-3}	
Li	0.53	Be	1.85
Na	0.97	Mg	1.74
K	0.86	Ca	1.54
		Sr	2.60
		Ba	3.51

Densities for comparison:
H_2O, 1.00 g cm^{-3}; Fe: 7.86 g cm^{-3}

Group 2

- They have reasonably high melting and boiling points.
- They have low densities although not as low as those in Group 1.
- They have colourless compounds.

Melting points show a general decrease down Group 1 and Group 2:

- the atoms increase in size
- the outer electrons are further away from the nucleus
- the attractive force of nuclei on the outer electrons decreases.

Melting points of s-block elements			
Group 1 Melting point/°C		Group 2 Melting point/°C	
Li	181		
Na	98	Mg	649
K	63	Ca	839
		Sr	769
		Ba	725

Melting points for comparison:
Fe: 1535°C; Cu 1083°C

Flame tests

EDEXCEL M2 SALTERS M1
CCEA M1, M2 WJEC M2

When heated in a flame, compounds of Groups 1 and 2 show characteristic flame colours, allowing different cations to be identified.

cation	Li+	Na+	K+	Ca2+	Ba2+	Sr2+
flame colour	red	yellow	lilac	brick-red	green	red

The characteristic flame colours arise from electron transitions within the atom (see pages 28–29).

> Heat energy from the flame excites an electron to a higher energy level.
> The electron returns to a lower energy level.
> The excess energy is emitted as light energy.
> The frequency of the flame colour depends upon the difference between the energy levels involved.
>
> **KEY POINT**

Reactivity of the s-block elements

AQA M2 OCR M1
EDEXCEL M2 SALTERS M1, M2
CCEA M2 WJEC M2

Li	Be
Na	Mg
K	Ca
Rb	Sr
Cs	Ba
Fr	Ra

reactivity **increases**

Down Group 1 and Group 2

atomic radius increases

first ionisation energy decreases

See page 73 for detailed explanation.

The elements in the s-block are the most reactive metals and are strong reducing agents.

Group 1 elements are oxidised in reactions, each atom losing one electron from its outer s sub-shell to form a 1+ ion (+1 oxidation state):

$$M \longrightarrow M^+ + e^-$$

Group 2 elements are oxidised in reactions, each atom losing two electrons from its outer s sub-shell to form a 2+ ion (+2 oxidation state):

$$M \longrightarrow M^{2+} + 2e^-$$

Reactivity **increases** down each group reflecting the increasing ease of losing electrons. The ionisation energy of the metal is an important factor in this process.

> Within each group, the elements become **more** reactive as the group descends.
> First ionisation energy **decreases** down the group.
>
> **KEY POINT**

The Group 2 elements

AQA M2 OCR M1
EDEXCEL M2 SALTERS M1
CCEA M2 WJEC M2

Reaction with oxygen

The Group 2 elements react vigorously with oxygen. Each element forms the expected ionic oxide with the general formula, $M^{2+}O^{2-}$:

e.g. $$2Ca(s) + O_2(g) \longrightarrow 2CaO(s)$$

Action of water

Reactions of Group 2 elements with oxygen and water are redox reactions.

Mg forms MgO with steam.

Reactivity increases down the group reflecting the increasing ease with which electrons can be lost.

- Mg reacts very slowly with water, forming the **hydroxide** and hydrogen:

$$Mg(s) + 2H_2O(l) \longrightarrow Mg(OH)_2(aq) + H_2(g)$$

- With steam, reaction is much quicker forming the **oxide** and hydrogen:

$$Mg(s) + H_2O(g) \longrightarrow MgO(s) + H_2(g)$$

- Further down the group from calcium, each metal reacts vigorously with water:

$$Ca(s) + 2H_2O(l) \longrightarrow Ca(OH)_2(aq) + H_2(g)$$

Group 2 oxides and hydroxides

$Mg(OH)_2$
$Ca(OH)_2$ solubility increases
$Sr(OH)_2$ alkalinity
$Ba(OH)_2$ increases

Thus solid barium hydroxide is reasonably soluble in water to form a strong alkaline solution with greater $OH^-(aq)$ concentration.

Reaction with water

The Group 2 oxides form alkaline solutions with water:

e.g. $$MgO(s) + H_2O(l) \longrightarrow Mg^{2+}(aq) + 2OH^-(aq)$$

- Magnesium hydroxide, $Mg(OH)_2(s)$ is only slightly soluble in water and so the resulting solution is only a dilute alkali.
- The solubility in water increases down the group and resulting solutions are more alkaline:

$$Ba(OH)_2(s) + aq \longrightarrow Ba^{2+}(aq) + 2OH^-(aq)$$

The alkalinity of the Group 2 hydroxides is exploited commercially. Some indigestion tablets contain $Mg(OH)_2$ to neutralise excess acid in the stomach. Farmers add $Ca(OH)_2$ as 'lime' to neutralise acid soils.

Reaction with acids

- The Group 2 oxides behave as bases and are neutralised by acids, such as $HCl(aq)$, forming salts and water:

e.g. $$MgO(s) + 2HCl(aq) \longrightarrow MgCl_2(aq) + H_2O(l)$$

Solubility of Group 2 sulfates (AQA and Edexcel only)

In medicine, barium sulfate is used as a 'barium meal' which shows up imperfections in the gut when exposed to X-rays. Barium compounds are extremely poisonous in solution but the insolubility of $BaSO_4$ is such that no harm is caused to the patient by this treatment.

Magnesium sulfate $MgSO_4$ is very soluble in water but the solubility decreases as the group is descended. The trend is so marked that barium sulfate $BaSO_4$ is virtually insoluble.

In the laboratory, the precipitation of $BaSO_4$ is used to test for the sulfate ion. A solution of a soluble barium salt (usually $BaCl_2(aq)$ or $Ba(NO_3)_2$) is added to a solution of a substance in dilute nitric acid. In the presence of aqueous sulfate ions, a dense white precipitate of barium sulfate is formed.

$$Ba^{2+}(aq) + SO_4^{2-}(aq) \longrightarrow BaSO_4(s)$$

$MgSO_4$
$CaSO_4$ solubility
$SrSO_4$ **decreases**
$BaSO_4$

The thermal stability of s-block compounds

| EDEXCEL | M2 | OCR | M1 |
| CCEA | M2 | SALTERS | M1 |

The s-block carbonates

Group 2 carbonates

$MgCO_3$
$CaCO_3$ ease of thermal
$SrCO_3$ decomposition
$BaCO_3$ **decreases**

The Group 2 carbonates are decomposed by heat:

e.g. $$MgCO_3(s) \longrightarrow MgO(s) + CO_2(g)$$

The Group 2 carbonates become more difficult to decompose as the group is descended.

Reason for trend (Edexcel and CCEA only)

The decomposition of the Group 2 carbonates is caused by the polarising effect of the Group 2 cation:

It is increasingly more difficult to decompose the carbonate as the group is descended. The polarising power of the metal cation decreases as the charge : size ratio decreases (see also pages 58–59).

> *Polarisation occurs when a small cation distorts the electron shells of a larger anion.*
>
> *The polarising effect is greatest between a small densely charged cation and a large anion.*

- As the size of the cation increases, the polarising effect becomes less.
- There is less weakening of the C–O bond.
- It becomes harder to decompose the carbonate.

$[Mg]^{2+}$

$[Ca]^{2+}$

$[Ba]^{2+}$

thermal stability of carbonates increases

increased ionic size gives a less concentrated charge and less polarisation of the carbonate anion

> *Group 2 hydroxides show a similar trend – thermal stability increases down the group.*

Group 1 carbonates (Edexcel only)

The cations in the Group 1 carbonates are larger than those of the Group 2 carbonates and their polarising power is less. The Li^+ ion, however, is comparable in size with the Mg^{2+} ion and lithium carbonate is also decomposed by heat:

$$Li_2CO_3(s) \longrightarrow Li_2O(s) + CO_2(g)$$

The remaining Group 1 carbonates are **not** decomposed by heat.

> *The Group 1 carbonates are thermally stable (except Li_2CO_3)*

The s-block nitrates (Edexcel only)

Group 2 nitrates

The Group 2 nitrates are decomposed by heat, releasing NO_2 and O_2 gases:

e.g. $$Ca(NO_3)_2(s) \longrightarrow CaO(s) + 2NO_2(g) + \tfrac{1}{2}O_2(g)$$

As with the Group 2 carbonates, the nitrates become more thermally stable as the group is descended. This again reflects the increasing size of the cations and the resultant decrease in polarising power.

Group 1 nitrates

The Group 1 nitrates decompose with heat releasing only oxygen gas:

e.g. $$NaNO_3(s) \longrightarrow NaNO_2(s) + \tfrac{1}{2}O_2(g)$$

However, as with the carbonates above, the increasing polarising power of the smaller Li^+ ion makes lithium nitrate less thermally stable and lithium nitrate is decomposed by heat in a similar way to the Group 2 nitrates:

$$2LiNO_3(s) \longrightarrow Li_2O(s) + 2NO_2(g) + \tfrac{1}{2}O_2(g)$$

> *With heat, Li_2CO_3 and $LiNO_3$ behave similarly to the carbonates and nitrates of magnesium. This is known as a diagonal relationship. You can see why by comparing the positions of Li and Mg in the Periodic Table.*

Progress check

1 Write down equations for the following reactions:
   (a) barium with water   (b) calcium oxide with nitric acid.
2 Identify the oxidation number changes taking place during the thermal decomposition of sodium nitrate:
   $$2NaNO_3(s) \longrightarrow 2NaNO_2(s) + O_2(g)$$

2 N, +5 ⟷ +3, O, −2 ⟷ 0.
(b) CaO(s) + 2HNO₃(aq) ⟶ Ca(NO₃)₂(aq) + H₂O(l)
1 (a) Ba(s) + 2H₂O(l) ⟶ Ba(OH)₂(aq) + H₂(g)

2.8 The Group 7 elements and their compounds

After studying this section you should be able to:

- describe the characteristic properties of the Group 7 elements
- recall the relative reactivity of the halogens as oxidising agents
- recall the characteristic tests for halide ions
- describe the characteristics of hydrogen halides
- describe the reactions of halides with concentrated sulfuric acid
- describe the use of thiosulfate titrations.

LEARNING SUMMARY

General properties

AQA	M2	OCR	M1
EDEXCEL	M2	SALTERS	M2
CCEA	M1	WJEC	M2

The common name for the elements in Group 7 is the **halogens**.

Electron configuration

Each halogen has **seven** outer shell electrons; just one electron short of the electron configuration of a noble gas. The outer **p** sub-shell contains **five** electrons.

Electron configuration of the halogens

F	[He] $2s^2 2p^5$
Cl	[Ne] $3s^2 3p^5$
Br	[Ar] $3d^{10} 4s^2 4p^5$
I	[Kr] $4d^{10} 5s^2 5p^5$
At	[Xe] $4f^{14} 5d^{10} 6p^2 6p^5$

F_2
Cl_2
Br_2
I_2

boiling point **increases** down group

Trend in physical states

The halogens exist as diatomic molecules, X_2. The boiling points of the halogens increase on descending the group.

- The physical states of the halogens at r.t.p. show the classic trend of gas $\longrightarrow$ liquid $\longrightarrow$ solid.
- On descending the group the number of electrons increases leading to an increase in van der Waals' forces between molecules.
- Therefore the boiling point increases.

Boiling points of the Halogens

	Boiling point/°C	State at r.t.p.
F_2	−188	gas
Cl_2	−35	gas
Br_2	59	liquid
I_2	184	solid
At_2	337	solid

Trend in electronegativity

| AQA | M2 |
| CCEA | M1 |

Electronegativity is a measure of the attraction of an atom for the pair of electrons in a covalent bond (see also page 62).

The hydrogen halides have polar molecules: $H^{\delta+}$–$X^{\delta-}$. The polarity decreases on descending the halogens and the order of polarity is:

most polar   H–F > H–Cl > H–Br > H–I   *least polar*

This trend in polarity results from the **decreasing** electronegativity of the halogen atom on descending the group.

- The atomic radius increases from F → Cl → I resulting in less nuclear attraction on the bonding electrons (this is despite the increase in nuclear charge!).
- There are more electron shells between the nucleus and the bonding electrons to shield the nuclear charge.

electronegativity of halogen **decreases**

polarity of H–X bond **decreases**

Electronegativity of the Halogens

	Pauling value
F	4.0
Cl	3.0
Br	2.8
I	2.5

The overall effect is that the smaller the halogen atom, the greater the nuclear attraction experienced by the bonding electrons. Thus the large electronegativity of fluorine results in a bond between hydrogen and fluorine that is more polar than between hydrogen and other halogens.

The relative reactivity of the halogens as oxidising agents

AQA	M2	OCR	M1
EDEXCEL	M2	SALTERS	M2
CCEA	M1	WJEC	M2

The halogens are the most reactive non-metals and are strong oxidising agents.

The halogens become **less** reactive as the group descends as their oxidising power decreases.

The halogens are reduced in reactions, each atom gaining one electron into a p sub-shell to form a –1 ion (–1 oxidation state):

e.g.
$$\frac{1}{2}F_2(g) \quad + \quad e^- \quad \longrightarrow \quad F^-(g)$$
$$[He]\ 2s^2 2p^5 \qquad\qquad\qquad [He]\ 2s^2 2p^6\ (or\ [Ne])$$

- An extra electron is captured by being attracted to the outer shell of an atom by the nuclear charge of an atom.

F
Cl
Br
I
At

→ reactivity as oxidising agent **decreases**

On descending the halogens:

- the atomic radii increase resulting in less nuclear attraction at the edge of the atom (despite the increase in nuclear charge).
- there are more electron shells between the nucleus and the edge of the atom to shield the nuclear charge.

F

Cl

Br

The overall effect is that most nuclear attraction is experienced at the edge of the small fluorine atoms. This attraction decreases down the group as the atoms get bigger.

Therefore the oxidising power of the halogens decreases down the group. Fluorine is the strongest oxidising agent and is able to attract an extra electron more strongly than other halogens.

Displacement reactions of the halogens

The decrease in reactivity as the group is descended can be demonstrated by displacement reactions of aqueous halides using Cl_2, Br_2 and I_2.

chlorine oxidises both Br^- and I^-:
$$Cl_2(aq) + 2Br^-(aq) \longrightarrow 2Cl^-(aq) + Br_2(aq)$$
$$Cl_2(aq) + 2I^-(aq) \longrightarrow 2Cl^-(aq) + I_2(aq)$$

bromine oxidises I^- only:
$$Br_2(aq) + 2I^-(aq) \longrightarrow 2Br^-(aq) + I_2(aq)$$

iodine does not oxidise either Cl^- or Br^-

The formation of halogens in displacement reactions is identified by colours, which are more distinctive in organic solvents:

halogen	water	hexane
Cl_2	pale green	pale green
Br_2	orange	orange
I_2	brown	purple

Industrial extraction of bromine

The main source of bromine is as bromide ions, Br^- in sea water. Bromine is extracted by oxidising sea water with chlorine. Because chlorine is a stronger oxidising agent, it displaces the bromide ions using the principles of the displacement reaction.

Reactions of halogens with metals and ions

Reaction of halogens with non-metals (CCEA only)

The halogens react readily with many non-metals.

Chlorine reacts explosively with hydrogen in the presence of sunlight.

$$H_2(g) + Cl_2(g) \longrightarrow 2HCl(g)$$

With phosphorus, trihalides are formed.

$$P_4(s) + 6Cl_2(g) \longrightarrow 4PCl_3(l)$$

Reaction of halogens with metals (Edexcel and CCEA only)

The halogens oxidise many metals forming ionic chlorides.

For example, sodium and magnesium are oxidised when heated with halogens.

$$2Na(s) + Cl_2(g) \longrightarrow 2NaCl(s)$$
$$Mg(s) + Cl_2(g) \longrightarrow MgCl_2(s)$$

Reaction of halogens with ions (Edexcel and CCEA only)

The halogens oxidise some ions to higher oxidation states.

For example, chlorine oxidises aqueous Fe(II) ions to Fe(III) ions.

$$2Fe^{2+}(g) + Cl_2(g) \longrightarrow 2Fe^{3+}(aq) + 2Cl^-(aq)$$

$$+2 \xrightarrow{\quad +1 \quad} +3$$
$$0 \xrightarrow{\quad -1 \quad} -1$$

Testing for halide ions

AQA	M2	OCR	M1
EDEXCEL	M2	SALTERS	M2
CCEA	M1	WJEC	M2

In sunlight, silver halides are reduced to silver. This reaction formed the basis of old photographic films.

Addition of aqueous silver ions (using $AgNO_3(aq)$) to a solution of halide ions in dilute nitric acid produces coloured precipitates that have different solubilities in aqueous ammonia.

fluoride: $Ag^+(aq) + F^-(aq) \xrightarrow{\quad\times\quad}$ no precipitate
chloride: $Ag^+(aq) + Cl^-(aq) \longrightarrow AgCl(s)$ white precipitate, soluble in dilute $NH_3(aq)$
bromide: $Ag^+(aq) + Br^-(aq) \longrightarrow AgBr(s)$ cream precipitate, soluble in conc. $NH_3(aq)$
iodide: $Ag^+(aq) + I^-(aq) \longrightarrow AgI(s)$ yellow precipitate, insoluble in conc. $NH_3(aq)$

Hydrogen halides

| EDEXCEL | M2 | CCEA | M1 |

Hydrogen halides react with ammonia gas to form ammonium salts:

$HCl(g) + NH_3(g) \rightarrow NH_4Cl(s)$

The hydrogen halides, HX, are colourless gases that are very soluble in water. They form strong acid solutions when dissolved in water. For example, hydrochloric acid forms when hydrogen chloride gas dissolves in water:

$$HCl(g) + aq \longrightarrow H^+(aq) + Cl^-(aq)$$

- As the group is descended, the H–X bond enthalpy decreases and the bond breaks more readily, resulting in an increase in acidity down the group.

The order of acidity is HI > HBr > HCl.

Reactions of halides with concentrated sulfuric acid

| AQA | M2 | CCEA | M1 |
| EDEXCEL | M2 |

HCl does **not** reduce H_2SO_4

Concentrated sulfuric acid is an oxidising agent. Halide salts react with concentrated sulfuric acid producing a range of products depending on the halide used. This difference is caused by the increasing reducing power of the hydrogen halides as the group is descended.

NaCl and H_2SO_4

Hydrogen chloride gas, HCl, is formed.

$$NaCl(s) + H_2SO_4(l) \longrightarrow NaHSO_4(s) + HCl(g)$$

The HCl formed is not a sufficiently strong reducing agent to reduce the sulfuric acid. No redox reaction takes place.

NaBr and H_2SO_4

Hydrogen bromide gas, HBr, is initially formed.

$$NaBr(s) + H_2SO_4(l) \longrightarrow NaHSO_4(s) + HBr(g)$$

Some of the hydrogen bromide reduces the sulfuric acid with the formation of sulfur dioxide and orange bromine fumes.

$$2HBr(g) + H_2SO_4(l) \longrightarrow SO_2(g) + Br_2(g) + 2H_2O(l)$$

$+6 \xrightarrow{-2} +4$ *S reduced*

$2 \times -1 \xrightarrow{+2} 0$ *Br oxidised*

> HBr reduces H_2SO_4
> $H_2SO_4 \xrightarrow{HBr} Br_2 + SO_2$

NaI and H_2SO_4

Hydrogen iodide gas, HI, is initially formed.

$$NaI(s) + H_2SO_4(l) \longrightarrow NaHSO_4(s) + HI(g)$$

Hydrogen iodide is a strong reducing agent and reduces the sulfuric acid to a mixture of reduced products including sulfur dioxide and hydrogen sulfide.

Reduction to SO_2 (ox no: +4)

$$2HI(g) + H_2SO_4(l) \longrightarrow SO_2(g) + I_2(s) + 2H_2O(l)$$

$+6 \xrightarrow{-2} +4$ *S reduced*

$2 \times -1 \xrightarrow{+2} 0$ *I oxidised*

Further reduction to H_2S (ox no: −2)

$$6HI(g) + SO_2(g) \longrightarrow H_2S(g) + 3I_2(s) + 2H_2O(l)$$

$+4 \xrightarrow{-6} -2$ *S reduced*

$6 \times -1 \xrightarrow{+6} 0$ *I oxidised*

> HI reduces H_2SO_4 and SO_2
> $H_2SO_4 \xrightarrow{HI} I_2 + SO_2$
> $SO_2 \xrightarrow{HI} I_2 + H_2S$

Disproportionation

AQA	M2	OCR	M1
EDEXCEL	M2	SALTERS	M2
CCEA	M1		

> **KEY POINT**
> **Disproportionation** is a reaction in which the same element is both oxidised and reduced.

Disproportionation of chlorine in water

Chlorine reacts with water. This reaction is an example of *disproportionation* in which chlorine is both reduced (to chloride, Cl^-) and oxidised (to chlorate(I) ClO^-).

$$Cl_2(aq) + H_2O(l) \longrightarrow HClO(aq) + HCl(aq)$$

$0 \xrightarrow{-1} -1$ *chlorine reduced*

$0 \xrightarrow{+1} +1$ *chlorine oxidised*

> A solution of chlorine in water is pale green, showing the presence of chlorine.

A small amount of chlorine is added to drinking water to kill bacteria that would make the water unsafe to drink. It has been claimed that chlorine treatment of water has done more to improve public health than any other treatment, preventing diseases such as cholera and typhoid. This is despite the toxicity of chlorine and possible risks from the formation of toxic chlorohydrocarbons by reaction with organic matter.

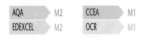

In some areas fluoride ions are added to drinking water to reduce tooth decay.

Some people believe that fluoride ions pose other detrimental health risks and that the rights of the individual to use or reject fluoride have been compromised.

Disproportionation of chlorine in dilute aqueous alkalis

Dilute aqueous sodium hydroxide reacts with halogens when mixed at room temperature.

This reaction is another example of *disproportionation* in which chlorine is both reduced (to chloride, Cl^-) and oxidised (to chlorate(I) ClO^-).

$$Cl_2(aq) + 2NaOH(aq) \longrightarrow NaCl(aq) + NaClO(aq) + H_2O(l)$$

$0 \xrightarrow{\ -1\ } -1$ *chlorine reduced*

$0 \xrightarrow{\ +1\ } +1$ *chlorine oxidised*

The solution formed is the basis for common bleach.

Disproportionation of chlorine in concentrated aqueous alkalis (Edexcel and CCEA only)

Further disproportionation of the halogen takes place with **hot, concentrated** sodium hydroxide.

Iodine first undergoes disproportionation into iodide, I^- and iodate(I), IO^-.

$$I_2(aq) + 2NaOH(aq) \longrightarrow NaI(aq) + NaIO(aq) + H_2O(l)$$

Iodate(I) ions then undergo further disproportionation.

$$3NaIO(aq) \longrightarrow 2NaI(aq) + NaIO_3(aq)$$

$2 \times +1 \xrightarrow{\ -4\ } 2 \times -1$ *iodine reduced*

$+1 \xrightarrow{\ +4\ } +5$ *iodine oxidised*

Halogens from electrolysis

SALTERS   M2

Halogens can be obtained by electrolysis of aqueous halide ions.

Halide ions are **oxidised**, discharging the halogen at the anode (positive electrode).

Electrolysis of brine (aqueous sodium chloride) is used for the industrial preparation of chlorine:

$$2Cl^-(aq) \longrightarrow Cl_2(g) + 2e^-$$

Iodine–thiosulfate titrations

EDEXCEL   M2

Starch is a sensitive test for the presence of iodine. If it is added too early in this titration, the deep-blue colour perseveres and the end-point cannot be detected with accuracy.

Iodine-thiosulfate titrations can be used to determine iodine concentrations, either directly or by liberating iodine using a more powerful oxidising agent.

Iodine oxidises thiosulfate ions:

$$I_2(aq) + 2S_2O_3^{2-}(aq) \longrightarrow 2I^-(aq) + S_4O_6^{2-}(aq)$$

- Thiosulfate is added from the burette to the iodine until the deep-brown colour of iodine becomes a light-straw colour.
- Starch is now added, forming a deep-blue colour.
- Thiosulfate is added dropwise until the deep-blue colour becomes colourless.
- This is the end-point and shows that all the iodine has just been reacted.

The estimation of chlorine in a bleach

10.0 cm³ of a sample of bleach was diluted to 250 cm³ with water. 25.0 cm³ of this solution was pipetted into a flask. An excess of aqueous potassium iodide was then added to liberate iodine:

equation 1 $Cl_2(aq) + 2I^-(aq) \longrightarrow I_2(aq) + 2Cl^-(aq)$

Titration of this solution required 23.2 cm³ of 0.100 mol dm⁻³ sodium thiosulfate.

equation 2 $2S_2O_3^{2-}(aq) + I_2(aq) \longrightarrow 2I^-(aq) + S_4O_6^{2-}(aq)$

Calculate the concentration of chlorine in the bleach.

From the titration results, amount (in moles) of Na₂S₂O₃ can be calculated:

amount (in mol) of $Na_2S_2O_3 = c \times \dfrac{V}{1000} = 0.100 \times \dfrac{23.2}{1000} = 0.00232$ mol

From the equations, the amount (in moles) of Cl₂ can be determined:

From *equation 2*, 2 mol $S_2O_3^{2-}$ reacts with 1 mol I_2

∴ 0.00232 mol $S_2O_3^{2-}$ reacts with $\dfrac{0.00232}{2}$ mol I_2 = 0.00116 mol I_2

From *equation 1*, 1 mol Cl_2 forms 1 mol I_2

∴ amount (in mol) of Cl_2 = 0.00116 mol

The concentration in mol dm⁻³ of the diluted bleach solution can now be calculated by scaling to 1000 cm³:

25.0 cm³ of diluted bleach contains 0.00116 mol Cl_2.

1 cm³ of diluted bleach contains $\dfrac{0.00116}{25}$ mol Cl_2.

1 dm³ of diluted bleach contains $\dfrac{0.00116}{25} \times 1000 = 0.0464$ mol Cl_2.

Finally, take into account the dilution with water from 10 cm³ to 250 cm³:

The original bleach was diluted by a factor of $\dfrac{250}{10} = 25$

∴ concentration of Cl_2 in the bleach is 25 x 0.0464 = 1.16 mol dm⁻³.

Compare acid-base titrations (page 40). The essential method is the same.

The concentration and volume of Na₂S₂O₃ are known.

Using the equations, determine the number of moles of Cl₂.

Work out the concentration of Cl₂(aq), in mol dm⁻³.

Don't forget to take into account any dilution.

Progress check

1 State and explain the trend in boiling points of the halogens fluorine to iodine.

2 How could you distinguish between NaCl, NaBr and NaI by a simple test?

3 Comment on the changes in oxidation number of chlorine in the following reaction: $Cl_2(aq) + H_2O(l) \longrightarrow HClO(aq) + H^+(aq) + Cl^-(aq)$

3 $Cl_2 \longrightarrow HClO$, Cl: 0 $\longrightarrow$ +1 (oxidation). $Cl_2 \longrightarrow Cl^-$, Cl: 0 $\longrightarrow$ −1 (reduction). Chlorine has been both oxidised and reduced (disproportionation).

2 Add AgNO₃(aq). NaCl gives a white precipitate, soluble in dilute ammonia. NaBr gives a cream precipitate, soluble in concentrated ammonia. NaI gives a yellow precipitate, insoluble in concentrated ammonia.

1 Boiling points increase down the group. From F_2 to I_2 the number of electrons increase leading to greater van der Waals' forces between molecules and higher boiling points.

2.9 Extraction of metals

After studying this section you should be able to:

- *understand that carbon reduction is used to extract iron from its ore*
- *understand that electrolysis is used to extract aluminium from its ore*
- *understand that metal reduction is used to extract titanium from its ore*
- *understand the importance of economic factors and recycling*

Reduction of metals

AQA ▸ M2

The main methods for extracting metals from their ores are:

- reduction of the ore with carbon
- reduction of the molten ore by electrolysis
- reduction of the ore with a more reactive metal.

Reduction of metal oxides with carbon

AQA ▸ M2

Production of iron

The main ore of iron, haematite, contains Fe_2O_3 and this is reduced by coke in the blast furnace. This is a continuous process in which high quality haematite, coke and limestone are fed in at the top of the furnace and hot air blown in near the bottom.

> Coke contains a very high carbon content. Coke is cheap and provides an economic source of carbon and carbon dioxide as reducing agents.

> The production of iron in the blast furnace is a **continuous** process. Raw materials are added at the top of the furnace and the products removed at the base. For further production, more raw materials are added and the process continues in a continuous flow, sometimes for months on end.

- Initially, hot coke reacts with air forming carbon monoxide:

$$2C(s) + O_2(g) \longrightarrow 2CO(g)$$

- The carbon monoxide reduces most of the iron oxide at around 1200°C:

$$Fe_2O_3(s) + 3CO(g) \longrightarrow 2Fe(l) + 3CO_2(g)$$

> Both C and CO are reductants in the blast furnace.

- In hotter parts of the furnace, coke also reacts directly with iron oxide:

$$2Fe_2O_3(s) + 3C(s) \longrightarrow 4Fe(l) + 3CO_2(g)$$

This equation represents the overall reaction for the reduction.

Limestone removes SiO_2.

The limestone is used to remove acidic impurities such as silicon dioxide (sand). This produces a slag, largely of calcium silicate, $CaSiO_3$:

$$CaCO_3 + SiO_2 \longrightarrow CaSiO_3 + CO_2$$

- Molten iron collects at the bottom of the furnace and is run off as 'pig iron.'
- Pig iron is very impure and brittle, containing about 4% of carbon (as well as Mn, Si, P and S).

Production of other metals

Reduction with carbon is a common method for obtaining many other metals from their ores.

- Manganese

  Pyrolusite ore contains manganese(IV) oxide. The ore is heated with carbon to obtain manganese by reduction:

$$MnO_2 + 2C \longrightarrow Mn + 2CO$$
$$MnO_2 + 2CO \longrightarrow Mn + 2CO_2$$

- Copper

  Copper is usually reduced from its sulfide ores. It can also be obtained by high temperature reduction using carbon:

$$CuO + C \longrightarrow Cu + CO$$
$$CuO + CO \longrightarrow Cu + CO_2$$

Environmental pollution problems:

sulfide ores form sulfur dioxide;

reduction with carbon produces greenhouse gases.

Aluminium oxide is very stable and could only be reduced by carbon at very high temperatures. Electrolysis is used instead to extract aluminium (see page 88).

Limitations of carbon reduction

Metal ores often contain metal oxides or sulfides.

- Sulfide ores (such as galena, PbS, and sphalerite, ZnS) are first roasted in air to produce the metal oxide. The sulfur is oxidised as sulfur dioxide, which contributes to acid rain. The sulfur dioxide can be reacted to make sulfuric acid, from which a wide variety of useful products can be manufactured.
- Oxide ores (such as haematite, Fe_2O_3) are reduced to the metal using fossil fuels such as coke from coal. Carbon dioxide, emitted as the main gaseous product, is a greenhouse gas.

Ores of some metals (such as titanium and tungsten) react with carbon forming metal carbides, so this is not a practical method for extracting these metals.

$$TiO_2 + 3C \longrightarrow TiC + 2CO$$

Reduction of metal oxides by electrolysis of the molten ore

AQA   M2

For metals more reactive than zinc, reduction with carbon does not take place except at very high temperatures and these metals are usually extracted by electrolysis of the molten ore.

The extraction of aluminium

The melting point of aluminium oxide is 2045°C but this is decreased using molten cryolite. This reduces the energy requirements.

The main ore of aluminium, bauxite, contains Al_2O_3. Purified bauxite is dissolved in molten cryolite (Na_3AlF_6) at 970°C. This is a continuous process needing regular additions of aluminium oxide.

The electrodes are made of graphite and the cell reactions are:

at the cathode $Al^{3+} + 3e^- \longrightarrow Al$
at the anode $2O^{2-} \longrightarrow O_2 + 4e^-$

- Periodically the graphite anodes need replacing because, at the high temperatures used, the carbon anodes react with the liberated oxygen.
$$2C + O_2 \longrightarrow 2CO$$
$$\text{and } C + O_2 \longrightarrow CO_2$$
- The process consumes large amounts of electricity (as electricity is needed to melt the Al_2O_3 as well as to reduce it).
- The process is only economic where electricity is relatively inexpensive.

Reduction of metal halides with metals

AQA   M2

Some metals, such as titanium, become brittle if contaminated with traces of impurities such as carbon, oxygen or nitrogen.

Highly pure metals can be formed by reduction of a metal halide with a reactive metal.

Despite the high cost, pure titanium is manufactured by reduction of its molten chloride with a more reactive metal.

When pure, titanium is a light metal with a high strength and high resistance to corrosion.

When impure, it is brittle and of little use.

Production of titanium

The main ore of titanium, rutile, contains titanium(IV) oxide. Titanium is manufactured in a two-stage batch process.

- Rutile is first converted to titanium(IV) chloride using chlorine and coke at around 900°C:
$$TiO_2 + 2C + 2Cl_2 \longrightarrow TiCl_4 + 2CO$$
- The titanium(IV) chloride is purified from other chlorides (e.g. those of iron, silicon and chromium) by fractional distillation under argon or nitrogen.
- The chloride is then reduced by a more reactive metal such as sodium or magnesium:

The production of titanium is a **batch** process. Raw materials are reacted together in a single 'batch' and the products are separated from the reaction mixture. For further production, the whole process needs to be repeated from scratch.

$$TiCl_4 + 4Na \longrightarrow Ti + 4NaCl$$
$$TiCl_4 + 2Mg \longrightarrow Ti + 2MgCl_2$$

- An inert atmosphere of argon is used to prevent any contamination of the metal with oxygen or nitrogen.
- If sodium has been used, the sodium chloride by-product is washed out with dilute hydrochloric acid, leaving titanium as a granular powder.
- If magnesium has been used, the magnesium chloride by-product is removed by distillation at high temperature and low pressure.

Reduction of metal oxides with hydrogen

AQA ▷ M2

The main ore of tungsten wolframite, is processed to precipitate tungsten(VI) oxide, WO_3. The tungsten(VI) oxide is reduced with hydrogen gas:

$$WO_3 + 3H_2 \longrightarrow W + 3H_2O$$

Hydrogen is more difficult to store than carbon and is highly flammable.

Economic factors and recycling

AQA ▷ M2

Costs of extracting a metal

The method used to reduce a metal on an industrial scale is dependent upon several factors:

The cost of the reducing agent

A reducing agent that is naturally available such as carbon (from coal) is cheaper than one such as sodium, which has to be prepared first by a separate (and often costly) process.

The energy costs for the process

The lower the temperature of a process, the lower the energy requirement and the cheaper the process. Carbon reduction is a high temperature process with substantial energy costs.

Instead of using high temperatures to extract copper, scrap iron can be added to solutions containing copper ions. The more reactive iron displaces copper from solution.

This process also allows copper to be extracted from low grade ore, containing small proportions of copper.

The purity required of the metal

It is relatively expensive to produce a metal with very high purity. There needs to be a high demand for the metal if this extra expense is to be justified.

These factors need to be weighed against each other when considering the overall cost of an extraction process.

Recycling

Metals are a valuable resource and, instead of disposal, metals are often collected and recycled.

Recycling of iron

Recycling metals conserves metal ores and saves energy.

- Iron is recycled by collection of scrap iron which is melted and reused.
- The Earth's reserves of iron ore are conserved.
- The energy required to mine iron ore, transport it and smelt it is several times greater than the energy required to recycle scrap iron.

Recycling of aluminium

- Owing to its very high resistance to corrosion, used aluminium is as good as new.
- Recycling conserves the Earth's supply of aluminium.
- The electrolysis of aluminium oxide requires high temperatures and vast quantities of electricity.

The cost of re-using scrap aluminium is only one-twentieth of the cost of making the pure metal.

- It is also cheaper and easier to refine recycled aluminium than to mine bauxite and extract the metal by electrolysis.

2.10 Qualitative analysis

After studying this section you should be able to:

- describe chemical tests for gases
- describe flame colours to identify metal ions
- describe tests for identifying and distinguishing between cations

LEARNING SUMMARY

Using chemical tests to identify ions

CCEA ▷ M2

Gas tests

gas	test
H_2	mild pop with lighted splint
O_2	relights glowing splint
Cl_2	turns moist indicator paper red, then bleached white
CO_2	turns lime water cloudy
SO_2	turns acidified dichromate paper from orange to green
HCl	forms dense white fumes with NH_3 fumes (from conc. NH_3(aq))
NH_3	forms dense white fumes with HCl fumes (from conc. HCl(aq))

Flame colours

Many potassium compounds contain traces of sodium compounds. The intense yellow Na^+ colour conceals the lilac K^+ colour.

Observing through blue glass filters out sodium's yellow colour, allowing potassium's lilac colour to be seen.

Flame colours can be used to identify some metal cations.

cation	Li^+	Na^+	K^+	Ca^{2+}	Ba^{2+}	Cu^{2+}
flame colour	red	yellow	lilac	brick-red	green	blue / green

Anion tests

SO_4^{2-}(aq) test

- Acidify solution with dilute nitric acid.
- Add aqueous barium chloride, $BaCl_2$(aq).
- SO_4^{2-} produces a white precipitate of barium sulfate, $BaSO_4$(s).

Distinguishing between CO_3^{2-}(aq) and HCO_3^-(aq)

- Add aqueous magnesium sulfate, $MgSO_4$(aq).
- CO_3^{2-} produces a white precipitate of $MgCO_3$(s).
- HCO_3^- produces no precipitate.

Tests for aqueous halide ions

See page 82 for more details of the reactions.

- Acidify with dilute nitric acid.
- Add aqueous silver nitrate, $AgNO_3$(aq).

halide	precipitate	solubility in NH_3(aq)
Cl^-	white precipitate of AgCl(s)	soluble in dilute NH_3(aq)
Br^-	cream precipitate of AgBr(s)	soluble in conc. NH_3(aq)
I^-	yellow precipitate of AgI(s)	insoluble in conc. NH_3(aq)

Cation tests

Ba^{2+}(aq) test

- Acidify with dilute nitric acid.
- Add aqueous potassium chromate, K_2CrO_4(aq).
- Ba^{2+} produces a deep-yellow precipitate of barium chromate, $BaCrO_4$(s).

Cu^{2+}(aq) test

- Add an excess of aqueous ammonia, NH_3(aq).
- Cu^{2+} produces a deep-blue solution.

NH$_4^+$(aq) test

- Heat with aqueous sodium hydroxide, NaOH(aq).
- NH_4^+ produces alkaline NH_3(g), which turns damp indicator paper blue.

Distinguishing between Fe^{2+}(aq) and Fe^{3+}(aq)

- Add aqueous sodium hydroxide, NaOH(aq).
- Fe^{2+} produces a pale-green precipitate of iron(II) hydroxide, $Fe(OH)_2$(s).
- Fe^{3+} produces an orange-brown precipitate of iron(III) hydroxide, $Fe(OH)_3$(s).

Distinguishing between Al^{3+}(aq), Mg^{2+}(aq) and Zn^{2+}(aq)

- Add a few drops of aqueous NaOH(aq) and NH_3(aq) separately.
- All three ions form white precipitates.
- Ions are distinguished by solubility in excess NaOH(aq) and NH_3(aq).

Notice that results from both NaOH(aq) and NH_3(aq) are needed to distinguish between the three ions.

	NaOH(aq)		*NH$_3$(aq)*	
	precipitate with few drops	*solubility in excess*	*precipitate with few drops*	*solubility in excess*
Al^{3+}(aq)	white	dissolves	white	insoluble
Mg^{2+}(aq)	white	**insoluble**	white	insoluble
Zn^{2+}(aq)	white	dissolves	white	**dissolves**

Fe^{3+}(aq) test

- Add aqueous potassium thiocyanate, KSCN(aq).
- Fe^{3+} produces a deep-blood-red solution.

Sample question and model answer

Chemical bonding helps to explain different properties of materials.

(a) Using suitable diagrams, explain what is meant by *ionic*, *covalent* and *metallic* bonding.

> An ionic bond is the electrostatic attraction between ions. ✓
> An example is sodium chloride:

> ✓ (dot and cross)
> ✓ (charges on ions)

> A covalent bond is a **shared** ✓ **pair** of electrons. ✓
> An example is hydrogen chloride.

> ✓ (dot and cross)

> Metallic bonding occurs between positive centres/ions ✓
> surrounded by mobile or delocalised electrons: ✓

> $(+) - (+) - (+)$
> $(+) - (+) - (+) - (+)$
> $(+) - (+) - (+)$

> Metallic bonding is the attraction between positive ions and mobile electrons. ✓

The phrase 'electrostatic attraction between ions' is essential when describing an ionic bond.

DO use diagrams in your answers. Here, dot and cross diagrams are essential for showing ionic and covalent compounds.

The word 'pair' is essential when describing a covalent bond. Omit it and you risk losing the mark.

There are plenty of examples. Keep them simple and make sure that you choose correctly. Don't use NaCl as a covalent example – it is ionic!

Many candidates do actually make this mistake.

(b) Three substances have ionic, covalent and metallic bonding respectively. Compare and explain the electrical conductivity of the three materials.

> The ionic compound does not conduct electricity when solid ✓
> because the ions are fixed ✓ in a lattice.
> The ionic compound does conduct electricity when aqueous or molten ✓
> because the ions are mobile. ✓
> The covalent compound does not conduct ✓ at all because there are no free charge carriers ✓ (electrons or ions).
> The metallic compound does conduct electricity ✓ because the delocalised electrons are able to move ✓ across a potential difference.

> 17 marking points ⟶ maximum of [15]

Special care needed throughout with language. A-grade students will score all of these marks.

*Ionic bonding: positive ions and negative ions – **both** carry electricity.*

*Metallic bonding: positive ions and delocalised electrons – **only** the electrons move and carry electricity.*

Practice examination questions

1 (a) A water molecule, H_2O, is bonded by covalent bonds. What is meant by the term *covalent bond*? [2]

(b) Use the formation of the H_3O^+ ion from water to explain what is meant by a *dative covalent bond*. [2]

(c) State the bond angle in a water molecule and predict, with an explanation, the bond angle in an H_3O^+ ion. [4]

(d) Name the major force of attraction that exists between molecules in water and explain how this type of force arises. [3]

[Total: 11]

2 Describe how *electron pair repulsion* explains the shapes of simple molecules. Show, using diagrams, how electron pair repulsion determines the molecular shapes and bond angles in molecules of (a) SiH_4; (b) PH_3; (c) BCl_3. [13]

[Total: 13]

3 When liquid bromine, Br_2, is heated gently, it forms an orange–brown vapour. When a crystal of potassium chloride is heated to the same temperature, it does not change state.

(a) Name the type of bonding or force that occurs:

(i) between bromine atoms in a molecule of bromine

(ii) between bromine molecules in liquid bromine. [2]

(b) Explain why liquid bromine turns into a vapour when heated gently. [1]

(c) Explain, in terms of its bonding, why the crystal of potassium chloride does not melt or vaporise when heated gently. [2]

(d) Describe what happens to the particles in potassium chloride when the solid is heated above room temperature but below its melting point. [1]

(e) Suggest why much more energy is required to vaporise potassium chloride than to melt it. [2]

[Total: 8]

4 The boiling points of (a) water, (b) hydrogen chloride and (c) krypton are shown below.

liquid	H_2O	HCl	Kr
boiling point /°C	100	−85	−152

Suggest reasons for the different boiling points by considering the nature and strength of the intermolecular forces in each case.

[Total: 12]

Practice examination questions

5 In the Periodic Table, describe and explain the trend in the atomic radii of the elements in Period 3 and in Group 2. [Total: 8]

6 Barium and magnesium are elements in Group 2 of the Periodic Table.
  (a) Complete and balance the following equations:
    (i) $Ba(s) + H_2O(l) \longrightarrow$
    (ii) $Mg(s) + O_2(g) \longrightarrow$
    (iii) $Mg(s) + HCl(aq) \longrightarrow$ [3]

  (b) (i) Suggest why the reactivity of the Group 2 elements increases on descending the group.
    (ii) Name one reaction of Group 2 elements that illustrates this trend of increasing reactivity. [4]

  (c) (i) What is the property of magnesium oxide that makes it suitable for its use as a lining in some furnaces?
    (ii) Name **two** other Group 2 compounds, and state a use for each of them. [3]

[Total: 10]

7 This question is about the Group 7 elements: chlorine, bromine and iodine.
  (a) Describe and explain the trend in oxidising ability shown by the Group 7 elements. [5]

  (b) Chlorine reacts with hot concentrated sodium hydroxide as in the equation below.
    $$3Cl_2(g) + 6NaOH(aq) \longrightarrow 5NaCl(aq) + NaClO_3(aq) + 3H_2O(l)$$
    (i) Use changes in oxidation numbers to show that this is a redox reaction.
    (ii) Calculate the maximum mass of $NaClO_3$ that could be prepared from the reaction of 65 dm³ of chlorine with hot concentrated sodium hydroxide at room temperature and pressure.
    (iii) Chlorine reacts with **dilute** aqueous sodium hydroxide at room temperature. Write an equation for this reaction and state what the resulting solution could be used for. [9]

[Total: 14]

8 (a) (i) What does the term *electronegativity* mean?
    (ii) What is the trend in electronegativity of the halogens fluorine to iodine? Explain why this trend happens. [5]

  (b) State and explain the trend in volatility of the halogens fluorine to iodine. [3]

  (c) Aqueous bromine was added separately to aqueous solutions of potassium chloride and potassium iodide. Describe what would be observed and write equation(s) for any reaction(s) that take place. [4]

  (d) State and explain the trend in reactivity shown by the experiments in part (c). [2]

[Total: 14]

Chapter 3
Energetics, rates and equilibrium

The following topics are covered in this chapter:

- Enthalpy changes
- Determination of enthalpy changes
- Bond enthalpy
- Reaction rates
- Catalysis
- Chemical equilibrium

3.1 Enthalpy changes

After studying this section you should be able to:

- understand that reactions can be exothermic or endothermic
- construct a simple enthalpy profile diagram for a reaction
- explain and use the terms: standard conditions, enthalpy changes of reaction, formation and combustion

LEARNING SUMMARY

Energy out, energy in

AQA	M2	OCR	M2
EDEXCEL	M1	SALTERS	M1
CCEA	M2	WJEC	M1

Enthalpy, H, is the **heat energy** that is stored in a chemical system.

Enthalpy cannot be measured experimentally. However, an **enthalpy change** can be measured from the temperature change in a chemical reaction.

> An enthalpy change ΔH is the heat energy exchange with the surroundings at constant pressure.
>
> **KEY POINT**

The conservation of energy is an important principle in science. This is often summarised by the first law of thermodynamics.

The first law of thermodynamics underpins all of this section.

> **The first law of thermodynamics** states that energy may be exchanged between a chemical system and the surroundings but the *total* energy remains constant.
>
> **KEY POINT**

it can't be destroyed or created but it can change form. (energy)

Exothermic reactions

EXO – EXIT

During an <u>exo</u>thermic reaction, heat energy is **released** to the surroundings.

Any energy **loss** from the chemicals is balanced by the same energy **gain** to the surroundings, which rise in temperature.

exothermic

chemicals lose energy:
ΔH **–ve**

surroundings gain energy and rise in temperature:
ΔT **+ve**

Energy pathway diagram (reaction profile)

released to surroundings

Enthalpy, H — reactants — ΔH –ve — products — reaction pathway

Energy **loss** during reaction

*Energy gain by the surroundings (identified by a temperature **rise**: ΔT +ve)*

Chemists refer to surroundings as anything other than the reacting chemicals.

Surroundings often just means the water in which chemicals are dissolved.

> In an exothermic reaction, ΔH is negative:
> - heat is given out (**to** the surroundings)
> - the reacting chemicals lose energy.
>
> **KEY POINT**

– has less energy than endothermic.

Endothermic reactions

During an endothermic reaction, heat energy is **taken in** from the surroundings.

Any energy **gain** to the chemicals is provided by the same energy **loss** from the surroundings, which fall in temperature.

Energy pathway diagram (reaction profile)

endothermic

chemicals gain energy:
ΔH **+ve**

surroundings lose energy and fall in temperature:
ΔT **−ve**

Energy loss from the surroundings
(identified by a temperature **fall**: ΔT –ve)

> **KEY POINT**
>
> In an endothermic reaction, ΔH is positive:
> - heat is taken in (**from** the surroundings)
> - chemicals gain energy.

Standard enthalpy changes

AQA	M2	OCR	M2
EDEXCEL	M1	SALTERS	M1
CCEA	M2	WJEC	M1

Enthalpy changes have been measured for many reactions. Many are recorded in data books as *standard enthalpy changes* and these are discussed below.

Standard conditions

$\Delta H^{\ominus}$ refers to an enthalpy (H) change (Δ) under standard conditions ($^{\ominus}$).

Standard pressure is 100 kPa or 1 bar.

The former standard pressure of 101 kPa or 1 atmosphere is still quoted in many books.

> **KEY POINT**
>
> *Standard conditions* are:
> - a pressure of 100 kPa
> - a stated temperature: 298 K (25°C) is usually used
> - a concentration of 1 mol dm^{-3} (*for aqueous solutions*).
>
> A *standard state* is the physical state of a substance under standard conditions.

The standard state of water at 298 K and 100 kPa is a liquid.

Standard enthalpy change of reaction

Always use a chemical equation or an unambiguous definition with a stated enthalpy change.

> **KEY POINT**
>
> The *standard enthalpy change of reaction* $\Delta H^{\ominus}_r$ is the enthalpy change that accompanies a reaction in the molar quantities that are expressed in a chemical equation under standard conditions, all reactants and products being in their standard states.

The enthalpy change of reaction $\Delta H^{\ominus}_r$ depends upon the quantities shown in a chemical equation. $\Delta H^{\ominus}_r$ should always be quoted with an equation.

$\Delta H^{\ominus}_r$ only has a meaning with an equation.

$\Delta H^{\ominus}_r$ has units of kJ mol^{-1}.

mol^{-1} means 'for the amount (in moles) shown in the equation'.

For the reaction:

$$H_2(g) \quad + \quad \tfrac{1}{2}O_2(g) \longrightarrow H_2O(l) \qquad \Delta H^{\ominus}_r = -286 \text{ kJ mol}^{-1}$$
1 mol   ½ mol   1 mol

but with twice the quantities, there is twice the enthalpy change:

$$2H_2(g) \quad + \quad O_2(g) \longrightarrow 2H_2O(l) \qquad \Delta H^{\ominus}_r = -572 \text{ kJ mol}^{-1}$$
2 mol   1 mol   2 mol

Standard enthalpy change of combustion   ~substance + O_2~

The *standard enthalpy change of combustion* $\Delta H^{\ominus}_{c}$ is the enthalpy change that takes place when one mole of a substance reacts completely with oxygen under standard conditions, all reactants and products being in their standard states.

This means that complete combustion of 1 mole of $C_2H_4(g)$ releases 1411 kJ of heat energy to the surroundings at 298 K and 100 kPa.

e.g. (C_2H_4(g)) + 3O_2(g) ⟶ 2CO_2(g) + 2H_2O(l) $\Delta H^{\ominus}_{c} = -1411$ kJ mol^{-1}

~1 mole~

$$C_2H_4(g) + 3O_2(g) \longrightarrow 2CO_2(g) + 2H_2O(l) \qquad \Delta H^{\ominus}_{c} = -1411 \text{ kJ mol}^{-1}$$

Standard enthalpy change of formation   ~elements → compound~

The *standard enthalpy change of formation* $\Delta H^{\ominus}_{f}$ is the enthalpy change that takes place when one mole of a compound in its standard state is formed from its constituent elements in their standard states under standard conditions.

This means that the formation of 1 mole of $H_2O(l)$ from 1 mole of $H_2(g)$ and ½ mole of $O_2(g)$ releases 286 kJ of heat energy to the surroundings at 298 K and 100 kPa.

e.g. $H_2(g) + \tfrac{1}{2} O_2(g) \longrightarrow \Big(H_2O(l)\Big)$ ~1 mole~ . $\Delta H^{\ominus}_{f} = -286$ kJ mol^{-1}

For an element, the standard enthalpy change of formation is defined as zero.

The formation of $H_2(g)$ from $H_2(g)$ does not involve a chemical change so there is no enthalpy change.

Standard enthalpy change of neutralisation (Edexcel and CCEA only)

Notice that
$\Delta H^{\ominus}_{f}(O_2) = 0$ kJ mol^{-1}
$\Delta H^{\ominus}_{r}$ for any element must be zero (formation of the element from the element – no change).

The *standard enthalpy change of neutralisation* $\Delta H^{\ominus}_{neut}$ is the energy change that accompanies the neutralisation of an acid by a base to form 1 mole of $H_2O(l)$, under standard conditions.

$$HCl(aq) + NaOH(aq) \longrightarrow NaCl(aq) + H_2O(l) \quad \Delta H^{\ominus}_{neut} = -57.9 \text{ kJ mol}^{-1}$$
acid   base   ⟶   1 mol

- In aqueous solution, strong acids and bases are completely dissociated and $\Delta H^{\ominus}_{neut}$ is approximately equal to the value above. This neutralisation process corresponds to the reaction: $H^+(aq) + OH^-(aq) \longrightarrow H_2O(l)$.
- For weak acids, this enthalpy change is less exothermic because some input of energy is required to ionise the acid (see also page 45).

Progress check

1 Draw energy profile diagrams for the following reactions:
   (a) $N_2O_4(g) \longrightarrow 2NO_2(g)$ $\Delta H = +58$ kJ mol^{-1}
   (b) $N_2(g) + 3H_2(g) \longrightarrow 2NH_3(g)$ $\Delta H = -92$ kJ mol^{-1}

2 Write an equation to represent the enthalpy change for:
   (a) $\Delta H^{\ominus}_{c}$ (CH$_4$);   (b) $\Delta H^{\ominus}_{f}$ (NO$_2$)

2 (a) $CH_4(g) + 2O_2(g) \longrightarrow CO_2(g) + 2H_2O(l)$
  (b) $\tfrac{1}{2}N_2(g) + O_2(g) \longrightarrow NO_2(g)$

3.2 Determination of enthalpy changes

After studying this section you should be able to:

- calculate enthalpy changes from direct experimental results, using the relationship: energy change = $mc\Delta T$
- use Hess' Law to construct enthalpy cycles
- determine enthalpy changes indirectly, using enthalpy cycles and enthalpy changes of formation and combustion

Direct determination of enthalpy changes

AQA	M2	OCR	M2
EDEXCEL	M1	SALTERS	M1
CCEA	M2	WJEC	M1

The **specific heat capacity** of a substance is the energy required to raise the temperature of 1 g of a substance by 1°C.

Assume that solids, such as Mg(s) make little difference to the energy change.

Any energy **gain** by the surroundings must have come from the same energy **loss** in the chemical reaction.

All values in this example are to 3 significant figures. This indicates the expected accuracy of the answer which should also be expressed to 3 significant figures.

In this case,
$\Delta H = -94.1$ kJ mol^{-1}.

Calculating enthalpy changes

The heat energy change, Q, in the surroundings can be calculated using the relationship below.

$$Q = mc\Delta T \text{ Joules}$$

- m is the mass of the surroundings that experience the temperature change
- c is the specific heat capacity of the surroundings
- ΔT is the temperature change (final temperature – initial temperature)

Example

Addition of an excess of magnesium to 100 cm³ of 2.00 mol dm^{-3} CuSO$_4$(aq) raised the temperature from 20.0°C to 65.0°C. Find the enthalpy change for the reaction:

$$Mg(s) + CuSO_4(aq) \longrightarrow MgSO_4(aq) + Cu(s)$$

specific heat capacity of solution $c = 4.18$ J g^{-1} K^{-1}
density of solution = 1.00 g cm^{-3}.

Find the heat energy change

100 cm³ of solution has a mass of 100 g;	
Temperature change, ΔT	= (65.0–20.0)°C
	= +45.0°C
Heat energy **gain** to surroundings, $Q = mc\Delta T$	= 100 × 4.18 × 45.0 J
	= **+18810 J**
∴ heat energy **loss** from the reacting chemicals	= **–18810 J**

Find out the amount (in moles) that reacted

Amount (in mol) of CuSO$_4$ that reacted = $2.00 \times \dfrac{100}{1000}$ mol = 0.200 mol

Scale the quantities to those in the equation

Mg(s) +	CuSO$_4$(aq)	$\longrightarrow$	MgSO$_4$(aq) +	Cu(s)
1 mol	1 mol		1 mol	1 mol

For 0.200 mol (1/5th mol) of CuSO$_4$, $\Delta H = -18810$ J
For 1 mol of CuSO$_4$, $\Delta H = 5 \times -18810 = -94050$ J
   $\Delta H = -94.1$ kJ mol^{-1} (to 3 sig. figs)

∴ enthalpy change of reaction is given by:

$$Mg(s) + CuSO_4(aq) \longrightarrow MgSO_4(aq) + Cu(s) \quad \Delta H = -94.1 \text{ kJ mol}^{-1}$$

Indirect determination of enthalpy changes

AQA	M2	OCR	M2
EDEXCEL	M1	SALTERS	M1
CCEA	M2	WJEC	M1

Hess' Law

Many reactions have enthalpy changes that cannot be found directly from a single experiment. Hess' Law provides a method for the *indirect* determination of enthalpy changes.

> Hess' Law is an extension of the First Law of Thermodynamics.

> **KEY POINT**
> *Hess' Law* states that, if a reaction can take place by more than one route and the initial and final conditions are the same, the total enthalpy change is the same for each route.

you take ~~enthalpy~~ *change no matter what route is the same!* ☺

The diagram below shows two routes for converting reactants into products.

hello chemistry learners!

enjoy reading this bit cnn. mwah. x.xx.
↓
" *Bored.*
f mL.

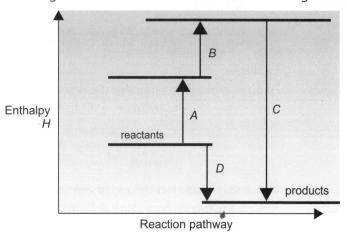

Following the arrows from reactants to products,
- *Route 1:* **A + B + C**
- *Route 2:* **D**

By Hess' Law, the total enthalpy change is the same for each route.
$$\therefore \ \mathbf{A + B + C = D}$$
If three of these enthalpy changes are known, the fourth can always be calculated.

> Indirect determination of an enthalpy change uses an energy cycle based on Hess' Law.

> This method is used when the reaction is very difficult to carry out and related reactions can be measured more easily.

Using $\Delta H^{\ominus}_c$ values to determine an enthalpy change indirectly

Example:

Find the enthalpy change for the reaction:
$$C(s) \ + \ 2H_2(g) \longrightarrow CH_4(g)$$

substance	C(s)	H$_2$(g)	CH$_4$(g)
$\Delta H^{\ominus}_c$ / kJ mol^{-1}	−394	−286	−890

> Enthalpy changes of combustion are required for all reactants and products.

Use the $\Delta H^{\ominus}_c$ data as a 'link' to construct an energy cycle.
- An energy cycle is constructed by linking the reactants and products to their **combustion products**.
- Note the direction of the arrows **from** the reactants and products of the reaction **to** the common combustion products.

> Always show your working. In exams, you are rewarded for a good method.

> If you make one small slip in a calculation, you may only lose 1 mark provided that you show clear working.

> The common combustion products here are CO$_2$(g) and H$_2$O(l). You can include these in your cycle but they are not required and 'combustion products' have been used here.

= −966 + 890 = −76· or *[reactants − products]*
−966 − (−890)
= −76 kJ mol⁻¹

99

Calculate the unknown enthalpy change

By Hess' Law,

Route 1: $\Delta H + [(-890)]$

Route 2: $[(-394) + (2 \times -286)]$

$\therefore \underbrace{\Delta H + [(-890)]}_{\text{Route 1}} = \underbrace{[(-394) + (2 \times -286)]}_{\text{Route 2}}$

(handwritten: (reatants − proaucts).)

(handwritten: (I was right))

$\Delta H = [(-394) + (2 \times -286)] - [(-890)]$

— $\Delta H = \mathbf{-76 \ kJ \ mol^{-1}}$

$\therefore C(s) + 2H_2(g) \longrightarrow CH_4(g) \qquad \Delta H^{\ominus} = -76 \ kJ \ mol^{-1}$

> **KEY POINT**
>
> **For enthalpy changes of combustion data only,**
>
> $\Delta H = \Sigma \Delta H^{\ominus}_c \ (\text{reactants}) - \Sigma \Delta H^{\ominus}_c \ (\text{products})$

> Enthalpy changes of formation are required for all reactants and products that are compounds.
>
> For elements, $\Delta H^{\ominus}_f = 0$ (formation of the element from the element – no change).

Using $\Delta H^{\ominus}_f$ values to determine an enthalpy change indirectly

Example:

Find the enthalpy change for the reaction:

$$C_2H_6(g) + 3\tfrac{1}{2}O_2(g) \longrightarrow 2CO_2(g) + 3H_2O(l)$$

substance	$C_2H_6(g)$	$CO_2(g)$	$H_2O(l)$
$\Delta H^{\ominus}_f / \ kJ \ mol^{-1}$	−85	−394	−286

Use the $\Delta H^{\ominus}_f$ data as a 'link' to construct an energy cycle

• An energy cycle is constructed linking the reactants and products with their constituent elements.

• Note the direction of the arrows from the constituent elements to the reactants and products of the reaction.

> Notice that
>
> $\Delta H^{\ominus}(O_2) = 0 \ kJ \ mol^{-1}$
>
> This can be omitted from the cycle.

(handwritten: − O₂ is an element.
∴ no ΔH°f.
+85 + −1646 = ‾1561
or
products − reactants.
−1646−(−85) = −1561.)

ΔH

$C_2H_6(g) + 3\tfrac{1}{2}O_2(g)$ → $2CO_2(g) + 3H_2O(l)$

(handwritten: +85)

$-85 + 3\tfrac{1}{2} \times 0$ **ROUTE 1** $(2 \times -394) + (3 \times -286)$

$= -85$

elements in standard states

(handwritten: ROUTE 2 −788 + ‾858 = −1646)

Calculate the unknown enthalpy change

By Hess' Law,

Route 1: $[(-85) + (3\tfrac{1}{2} \times 0)] + \Delta H$

Route 2: $[(2 \times -394) + (3 \times -286)]$

$\therefore \underbrace{[(-85) + 0] + \Delta H}_{\text{Route 1}} = \underbrace{[(2 \times -394) + (3 \times -286)]}_{\text{Route 2}}$

$\therefore \Delta H = [(2 \times -394) + (3 \times -286)] - [(-85) + 0]$

$\therefore \Delta H = \mathbf{-1561 \ kJ \ mol^{-1}}$

$C_2H_6(g) + 3\tfrac{1}{2}O_2(g) \longrightarrow 2CO_2(g) + 3H_2O(l) \qquad \Delta H^{\ominus} = -1561 \ kJ \ mol^{-1}$

> **KEY POINT**
>
> **For enthalpy changes of formation data only,**
>
> $\Delta H = \Sigma \Delta H^{\ominus}_f \ (\text{products}) - \Sigma \Delta H^{\ominus}_f \ (\text{reactants})$

Progress check

For questions 1 and 2 assume that:

specific heat capacity of solution
$c = 4.18$ J g^{-1} K^{-1};

density of solution
$= 1.00$ g cm^{-3}.

1 Addition of zinc powder to 55.0 cm³ of aqueous copper(II) sulfate at 22.8 °C raised the temperature to 32.3 °C. 0.3175 g of copper were obtained.
(a) Calculate the energy released in this reaction.
(b) Write an equation, including state symbols, for this reaction.
(c) Calculate the enthalpy change for this reaction per mole of copper formed.

2 Combustion of 1.60 g of ethanol, C_2H_5OH, raised the temperature of 150 g of water from 22.0 °C to 71.0 °C. Find the enthalpy change of combustion of ethanol.

3 Use the $\Delta H^{\ominus}_c$ data below to calculate enthalpy changes for:
(a) $3C(s) + 4H_2(g) \longrightarrow C_3H_8(g)$
(b) $C(s) + 2H_2(g) + \frac{1}{2}O_2(g) \longrightarrow CH_3OH(l)$

substance	$\Delta H^{\ominus}_c$ / kJ mol^{-1}
C(s)	−394
$H_2(g)$	−286
$C_3H_8(g)$	−2219
$CH_3OH(l)$	−726

4 Use the $\Delta H^{\ominus}_f$ data below to calculate enthalpy changes for:
(a) $C_2H_4(g) + H_2(g) \longrightarrow C_2H_6(g)$
(b) $SO_2(g) + 2H_2S(g) \longrightarrow 2H_2O(l) + 3S(s)$

compound	$\Delta H^{\ominus}_f$ / kJ mol^{-1}
$C_2H_4(g)$	+52
$C_2H_6(g)$	−85
$SO_2(g)$	−297
$H_2S(g)$	−21
$H_2O(l)$	−286

4 (a) −137 kJ mol⁻¹; (b) −233 kJ mol⁻¹.
3 (a) −107 kJ mol⁻¹; (b) −240 kJ mol⁻¹.
2 −883 kJ mol⁻¹.
1 (a) 2.18 kJ; (b) Zn(s) + CuSO₄(aq) ⟶ Cu(s) + ZnSO₄(aq); (c) −437 kJ mol⁻¹.

55cm.

58 × 1.00 ← $\frac{55}{1000}$. = 0.0.55 moles .
            0.06 .

− 55 × 4.18 × 9.5 = $\frac{2184.05}{1000}$ = 2.18 KJ mol⁻¹ .

$\frac{0.3175.}{63.5.}$ = 5×10³·

1(

3.3 Bond enthalpy

After studying this section you should be able to:

- *understand and use the term bond enthalpy*
- *explain chemical reactions in terms of enthalpy changes associated with the breaking and making of chemical bonds*
- *determine enthalpy changes indirectly, using average bond enthalpies*

LEARNING SUMMARY

Bond enthalpy

AQA	M2	OCR	M2
EDEXCEL	M1	SALTERS	M1
CCEA	M2	WJEC	M1

| AQA (A2) | M4 |

Bond enthalpies are **positive** and refer to **bond breaking** – this process requires energy.

Enthalpy is stored within chemical bonds and **bond enthalpy** indicates the strength of a chemical bond in a gaseous molecule. For simple molecules, such as $H_2(g)$ and $HCl(g)$, bond enthalpy applies to the following processes:

$$H-H(g) \longrightarrow 2H(g) \qquad \Delta H = +436 \text{ kJ mol}^{-1}$$
$$H-Cl(g) \longrightarrow H(g) + Cl(g) \quad \Delta H = +432 \text{ kJ mol}^{-1}$$

> **Bond enthalpy** is the enthalpy change required to **break** and separate **1 mole of bonds** in the molecules of a gaseous element or compound so that the resulting gaseous species exert no forces upon each other.

KEY POINT

See also Activation Energy: page 105.

Average bond enthalpies

Bond enthalpies such as those above (H–H and H–Cl) apply to specific compounds. Only H_2 can have H–H bonds and the H–H bond enthalpy shown above has a definite value. However, some bonds can have different strengths in different environments.

Not all C–H bonds are created equal.

The Cl atom in CH_3Cl affects the environment of the C–H bonds.

C–H bonds have different strengths and different bond enthalpies

Data books provide an indication of the likely bond enthalpy of a particular bond by listing **average** or **mean bond enthalpies**.

An *average* bond enthalpy indicates the strength of a *typical* bond.
The average bond enthalpy for the C–H bond is +413 kJ mol^{-1}.

Bond making and bond breaking

Bond breaking requires energy: ENDOTHERMIC .
Bond making releases energy: EXOTHERMIC.

Chemical reactions involve bond breaking followed by bond making.

- Energy is first needed to break bonds in the reactants.
  Bond breaking is an endothermic process and **requires** energy.
- Energy is then released as new bonds are formed in the products.
  Bond making is an exothermic process and **releases** energy.

Bond enthalpy is an endothermic change (ΔH +ve) for bonds being broken.

When bonds are made, the enthalpy change will be the same magnitude but the opposite sign (ΔH –ve).

Using bond enthalpies to determine enthalpy changes

The enthalpy change for a reaction involving simple gaseous molecules can be determined using average bond enthalpies in an **energy cycle**:

- Enthalpy required to break bonds = Σ(bond enthalpies in reactants)
- Enthalpy released to make bonds = $-\Sigma$(bond enthalpies in products)

Notice that the relative strengths of the bonds in the reactants and the bonds in the products decide whether a reaction is exothermic or endothermic.

> **KEY POINT**
>
> $\Delta H = \Sigma$(bond enthalpies in reactants) $- \Sigma$(bond enthalpies in products).

For the reaction: $CH_4(g) + 2O_2(g) \longrightarrow CO_2(g) + 2H_2O(g)$,

$$H-\underset{\underset{H}{|}}{\overset{\overset{H}{|}}{C}}-H \ + \ \begin{matrix} O{=}O \\ O{=}O \end{matrix} \longrightarrow O{=}C{=}O \ + \ \begin{matrix} H\diagdown{}_O\diagup H \\ H\diagdown{}_O\diagup H \end{matrix}$$

	4 (C–H) + 2 (O=O)	2 (C=O) + 4 (O–H)
ΔH/kJ mol^{-1}	$(4 \times 413) + (2 \times 497)$	$(2 \times 805) + (4 \times 463)$
	Bonds broken: (endothermic)	Bonds made: (exothermic)

$\Delta H = \Sigma$(bond enthalpies in reactants) $- \Sigma$(bond enthalpies in products)
$\therefore \Delta H = [\,(4 \times 413) + (2 \times 497)\,] - [\,(2 \times 805) + (4 \times 463)\,]$ kJ mol^{-1}
$= \mathbf{-816}$ **kJ mol^{-1}**

Note that the calculated value is only approximate – the actual bond enthalpies involved may differ from the average values.

- Bonds with small bond enthalpies will break first.
- Low bond enthalpies indicate that a reaction will take place quickly.

Entropy

SALTERS   M1

Entropy is a measure of the number of ways that molecules and their associated energy quanta can be arranged.

In chemical reactions, energy changes depend not only on enthalpy but also on temperature and entropy.

Entropy, S, is a measure of the number of ways that molecules can be arranged.

Entropy increases whenever particles become more disordered, e.g.

- when a gas spreads spontaneously through a room
- with increasing temperature
- during changes in state: solid $\longrightarrow$ liquid $\longrightarrow$ gas
- when a solid lattice dissolves

$$NaCl(aq) + aq \longrightarrow Na^+(aq) + Cl^-(aq)$$

At 0 K, perfect crystals have zero entropy.

- in a reaction in which there is an increase in the number of gaseous molecules (i.e. when a gas is evolved).

$$NaHCO_3(s) + HCl(aq) \longrightarrow NaCl(aq) + H_2O(l) + CO_2(g)$$
$$MgCO_3(s) \longrightarrow MgO(s) + CO_2(g).$$

Progress check

1 Use the data in the table to find the enthalpy change for each reaction.
   (a) $C_2H_4(g) + 3O_2(g) \longrightarrow 2CO_2(g) + 2H_2O(g)$
   (b) $N_2(g) + 3H_2(g) \longrightarrow 2NH_3(g)$

bond	C–H	C=O	O=O	O–H	C=C	C–C	N≡N	H–H	N–H
bond average enthalpy /kJmol^{-1}	+413	+805	+497	+463	+612	+347	+945	+436	+391

1 (a) −1317 kJ mol^{-1}; (b) −93 kJ mol^{-1}.

3.4 Reaction rates

After studying this section you should be able to:

- *recall the factors which control the rate of a chemical reaction*
- *understand, in terms of collision theory, the effect of concentration changes on the rate of a reaction*
- *explain the significance of activation energy using the Boltzmann distribution*
- *use the Boltzmann distribution to explain the effect of temperature changes on a reaction rate*
- *understand that an enthalpy change indicates the feasibility of a reaction*

LEARNING SUMMARY

What is a reaction rate?

AQA	M2	OCR	M2
EDEXCEL	M2	SALTERS	M2
CCEA	M2	WJEC	M1

The rate of a chemical reaction is a measure of how quickly a reaction takes place.

> **KEY POINT**
>
> The rate of a reaction is usually measured as the rate of change of **concentration** of a stated species in a reaction.
>
> The units of rate are mol dm^{-3} s^{-1}.

Measuring reaction rates

Although concentration can be measured, other quantities proportional to concentration, such as gas volumes, may be easier to monitor. Rates of different reactions show a very wide variation and it may be more convenient to measure a reaction rate per minute or per hour rather than per second.

What affects a reaction rate?

The reasons why reaction rates are affected by these factors is discussed in the rest of this chapter.

The rate of a reaction may be affected by the following factors:

- the concentration of the reactants
- the surface area of solid reactants
- a temperature change
- the presence of a catalyst.

In some reactions, such as photosynthesis, the rate is affected by the presence and intensity of radiation.

How does reaction rate change during a reaction?

A is being used up – its concentration decreases.

B is being formed – its concentration increases.

For a reaction: **A** $\longrightarrow$ **B**, the reaction rate $= \dfrac{\text{change of concentration}}{\text{time}}$

The reaction rate can be measured as:

- the rate of **decrease** in concentration of **A**
- the rate of **increase** in concentration of **B**.

The graphs below show how the concentrations of **A** and **B** change during the course of the reaction: **A** $\longrightarrow$ **B**.

A is used up and its concentration falls. **B** is formed and its concentration increases.

- The steeper the gradient, the faster the rate of the reaction.
- The reaction is fastest at the start when the concentration of **A** is greatest.
- As the reaction proceeds, the rate slows down because the concentration of **A** decreases.
- When the reaction is complete, the graph levels off and the gradient becomes zero.

Activation energy

AQA	M2	OCR	M2
EDEXCEL	M2	SALTERS	M2
CCEA	M2	WJEC	M1

The activation energy of a reaction is the minimum energy required for the reaction to occur.

Only those collisions of sufficient energy to overcome the activation energy lead to a reaction.

In a gas or solution, particles are in constant motion and they collide with each other, with any solid species and with the walls of their container. When particles collide, a reaction can only take place if the energy of the collision exceeds the activation energy of the reaction.

The *activation energy* of a reaction is the energy required to start a reaction by breaking bonds (see page 102).

- Activation energy is often supplied by a spark or by heating the reactants.
- Reactions with a small activation energy often take place very readily.
- A large activation energy may 'protect' the reactants from taking part in the reaction (at room temperature).

The Boltzmann distribution

AQA	M2	OCR	M2
EDEXCEL	M2	SALTERS	M2
CCEA	M2	WJEC	M1

The Boltzmann, or Maxwell-Boltzmann, distribution shows the distribution of molecular energies in a gas at constant temperature.

The Boltzmann distribution

Characteristics of the Boltzmann distribution
- Most gas molecules have energies within a comparatively narrow range.
- The curve will only meet the energy axis at infinity energy. No molecules have zero energy.
- The area under the distribution curve gives the total number of gas molecules.
- Only those molecules with more energy than the activation energy of the reaction are able to react.

The effect of a concentration change on reaction rate

If the concentration of a reactant in solution or in a gas mixture is increased,
- there are more particles present per volume
- more collisions take place each second
- more collisions exceed the activation energy every second
- therefore the rate of reaction increases.

The diagram below shows distribution curves for two concentrations, C_1 and C_2. where concentration, C_2 > concentration, C_1.

- Providing the temperature is the **same**, distribution curves for different concentrations have the **same** shape.

> A reaction can only take place if the activation energy is exceeded.

> Note that the proportion of the total number of molecules exceeding the activation energy is the same. The rate increases because there are more molecules per volume and more molecules must now exceed the activation energy.

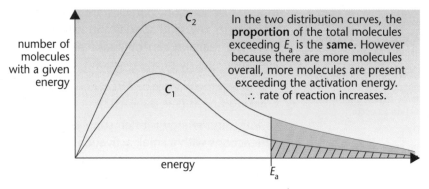

In the two distribution curves, the **proportion** of the total molecules exceeding E_a is the **same**. However because there are more molecules overall, more molecules are present exceeding the activation energy. ∴ rate of reaction increases.

For reactions involving gases, increasing the pressure also increases the concentration of any gas. This results in an increased reaction rate.

The effect of a change in surface area on reaction rate

For a reaction involving a solid, the reaction takes place at a faster rate when the solid is in a powdered form rather than as lumps.
Calcium carbonate reacts with hydrochloric acid producing carbon dioxide gas:

$$CaCO_3(s) + 2HCl(aq) \longrightarrow CaCl_2(aq) + H_2O(l) + CO_2(g)$$

By measuring the volume of carbon dioxide gas evolved with time, the rate of this reaction can be monitored. The graph below compares the reaction rates when using an excess of calcium carbonate as powder and as lumps. The same volume of the same concentration of hydrochloric acid has been reacted in both experiments.

> The gradient provides an indication of reaction rate.
>
> The steeper the curve, the faster the reaction.

> The final volume is the same in both experiments since the same amount (in mol) of hydrochloric acid has been reacted with an excess of calcium carbonate.

The gradient of the graph is much steeper with powdered carbonate because the surface area available for the reaction with hydrochloric acid is much greater than with lumps, allowing more collisions per second.

The effect of a temperature change on reaction rate

The average kinetic energy of the particles is proportional to temperature. As temperature increases, so does the kinetic energy of gas molecules.

The diagram below shows distribution curves for a sample of gas at two temperatures, T_1 and T_2, where temperature, T_2 > temperature, T_1.

Increasing the temperature **moves** the distribution curve to the right.

The Boltzmann distribution curve is displaced to the right with the peak lower. The average energy is now increased.

Increasing the temperature **does not change** the activation energy or the total number of molecules – the **shape of the curve changes**.

A **greater proportion** of molecules **exceeds the activation energy at higher temperature**.

The total area under each curve is a measure of the total number of molecules present, and this is the same for each curve.

An increase in temperature increases the rate of a reaction because:

- the molecules move faster and have more kinetic energy
- there are more collisions each second
- the increased kinetic energy produces more energetic collisions
- **a greater proportion of molecules exceed the activation energy.**

With even small increases in temperature, the shift in the distribution curve greatly increases the number of molecules exceeding the activation energy. This means that a small temperature rise can lead to a large increase in rate. Many reactions double their rate for each 10°C increase in temperature.

Progress check

1 Explain the following, in terms of collision theory and distribution graphs.
(a) Reactions take place quicker when the reactants are more concentrated.
(b) Reactions take place quicker at higher temperatures.

1 (a) Because there are more molecules per volume, more collisions take place each second. More collisions exceed the activation energy every second, increasing the reaction rate.
(b) The molecules move faster and have more kinetic energy. There are more collisions each second and the increased kinetic energy produces more energetic collisions. A greater proportion of molecules exceeds the activation energy.

3.5 Catalysis

After studying this section you should be able to:

- *understand what is meant by a catalyst*
- *explain how a catalyst changes the activation energy of a reaction*
- *explain what is meant by homogeneous and heterogeneous catalysis*

How do catalysts work?

AQA	M2	OCR	M2
EDEXCEL	M2	SALTERS	M1, M2
CCEA	M2	WJEC	M1

These key points are often tested in exams.

What is a catalyst?

A catalyst changes the rate of a chemical reaction, but is unchanged at the end of the reaction. Most catalysts speed up reaction rates although there are some (called inhibitors) that slow down reactions.

> **KEY POINT**
>
> A catalyst speeds up the rate of a reaction by providing an **alternative route** for the reaction with a **lower activation energy**.
> Activation energy with catalyst, E_c < Activation energy without catalyst, E_a

The effects of a catalyst on reaction route and activation energy are shown in the diagrams below.

Energy profile diagram

Notice the way the catalyst reduces the energy barrier to the reaction.

The catalyst provides an alternative route in which $E_c < E_a$

The catalyst does **not** change the distribution curve.

Notice how more molecules have an energy exceeding the new, lower activation energy.

Boltzmann distribution curve

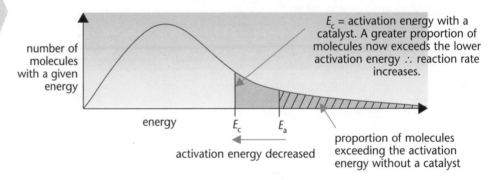

E_c = activation energy with a catalyst. A greater proportion of molecules now exceeds the lower activation energy ∴ reaction rate increases.

activation energy decreased

proportion of molecules exceeding the activation energy without a catalyst

Heterogeneous and homogeneous catalysis

| CCEA | M2 | SALTERS | M1, M2 |
| OCR | M2 | WJEC | M1 |

A **heterogeneous** catalyst is in a **different** phase from the reactants. Many reactions between **gases** are catalysed using a **solid** catalyst.

A **homogeneous** catalyst is in the **same** phase as the reactants. Many reactions between **gases** in the atmosphere are catalysed using a **gaseous** catalyst. Homogeneous catalysis also takes place in many **aqueous** reactions in which both the reactants and the catalyst are dissolved in water.

hetero- = different

homo- = same

Heterogeneous catalysis

Many heterogeneous catalysts are transition metals or their compounds.

The examples below are gaseous reactions catalysed by a solid catalyst.

Haber Process for ammonia (see also pages 113–115)

$$N_2(g) + 3H_2(g) \rightleftharpoons 2NH_3(g)$$

- The **gaseous** equilibrium process is catalysed by a **solid** iron catalyst, Fe(s).

surface of iron catalyst
Reactant molecules diffuse
towards the catalyst surface

Reactant molecules adsorbed at
active sites on the catalyst surface
Bonds weaken allowing reaction
to take place

Following reaction, the product
molecules diffuse away from the
catalyst surface

Polluting gases temporarily bond to the catalyst's surface and it is here that reactions are able to take place. The products then leave the surface, allowing processing of more polluting gases by the catalyst.

Contact process for sulfur trioxide (and subsequently sulfuric acid)

$$2SO_2(g) + O_2(g) \rightleftharpoons 2SO_3(g)$$

- The **gaseous** equilibrium process is catalysed by **solid** vanadium(V) oxide, V_2O_5(s).

Catalytic converters

Many polluting gases, such as carbon monoxide, CO, nitrogen monoxide, NO and unburnt hydrocarbons are produced in a car engine.

These gases are formed by:

The catalytic converter has a surface area equivalent to two football pitches! The reacting gases now have a much greater chance of being adsorbed onto the catalyst so that a reaction can take place.

- incomplete combustion of hydrocarbons forming CO
- reaction between N_2 and O_2 in a hot car engine forming NO.

These polluting **gases** are made to react using the catalytic converter in a car's exhaust system. The **solid** Rh/Pt/Pd catalyst is supported on a ceramic honeycomb which gives a large surface area for catalysis, a feature that greatly improves the effectiveness of many solid heterogeneous catalysts.

The catalytic converter removes much of this pollution in a series of reactions, e.g.

$$2CO(g) + 2NO(g) \longrightarrow 2CO_2(g) + N_2(g)$$

Homogeneous catalysis (not CCEA)

A homogeneous catalyst does react but it is then regenerated, to be used again in a chain reaction – the middleman of chemistry.

During homogeneous catalysis, the catalyst **does** react forming an *intermediate state* with **lower activation energy**. Subsequently, the catalyst is regenerated, allowing further reaction to take place. Overall the catalyst is not used up, but it is actively involved in the process.

The breakdown of atmospheric ozone, catalysed by Cl· radicals

The Cl· radicals are recycled in a chain reaction. A single Cl radical can break down thousands of O_3 molecules.

Chlorine radicals, Cl· are formed in the stratosphere by the action of ultraviolet radiation on chlorofluorocarbons, CFCs. The **gaseous** Cl· radicals catalyse the breakdown of **gaseous** ozone – the catalysis is homogeneous.

For further details of the role of Cl· radicals in ozone depletion see page 149.

chain reaction:
$$O_3(g) + Cl\cdot(g) \longrightarrow O_2(g) + ClO\cdot(g)$$
$$O(g) + ClO\cdot(g) \longrightarrow O_2(g) + Cl\cdot(g)$$

overall:
$$O_3(g) + O(g) \longrightarrow 2O_2(g)$$

The O(g) radicals are formed naturally in the stratosphere by the action of ultraviolet radiation on O_2(g).

See also radical substitution of alkanes, page 128.

3.6 Chemical equilibrium

After studying this section you should be able to:

- explain the main features of a dynamic equilibrium
- predict the effects of changes on the position of equilibrium
- know that a catalyst does not affect the position of equilibrium
- understand why a compromise temperature and pressure may be used to obtain an economic yield in industrial processes

Reversible reactions and dynamic equilibrium

AQA	M2	OCR	M2
EDEXCEL	M2	SALTERS	M2
CCEA	M2	WJEC	M1

Reversible reactions

Many chemical reactions take place until the reactants are completely used up and such reactions *'go to completion'*.

> In a chemical equation,
> *reactants* are on the left-hand side;
> *products* are on the right-hand side.

An example of a reaction that goes to completion is the oxidation of magnesium:

$$2Mg(s) + O_2(g) \longrightarrow 2MgO(s)$$

- The sign '$\longrightarrow$' indicates that the reaction takes place from left to right: this is called the *forward direction*.

Many reactions are **reversible**: they can take place in either direction.

An example of a reaction that is reversible is the formation of ammonia $NH_3(g)$ from nitrogen $N_2(g)$ and hydrogen $H_2(g)$.

$$N_2(g) + 3H_2(g) \rightleftharpoons 2NH_3(g)$$

- The sign '$\rightleftharpoons$' indicates that the reaction is reversible.

> The reversible reaction:
> $N_2(g) + 3H_2(g) \rightleftharpoons 2NH_3(g)$
> is used to illustrate dynamic equilibrium throughout this chapter.
>
> This is a homogeneous equilibrium – one in which all reactants and products have the same phase. In this system, all are gases.

Approaching equilibrium

- If nitrogen and hydrogen are mixed together, the reaction can proceed only in the forward direction because no products are yet present:

$$N_2(g) + 3H_2(g) \longrightarrow$$

- As the reaction proceeds, ammonia is formed. The ammonia starts to react in the *reverse direction*, indicated by '$\longleftarrow$', forming nitrogen and hydrogen. At first, there is so little ammonia present that the reverse process takes place extremely slowly:

$$N_2(g) + 3H_2(g) \rightleftharpoons 2NH_3(g)$$

- As more ammonia is formed the reverse process takes place faster. Nitrogen and hydrogen are being used up and the forward reaction slows down. Eventually, the reverse process takes place at the same rate as the forward reaction and **equilibrium** is attained:

$$N_2(g) + 3H_2(g) \rightleftharpoons 2NH_3(g)$$

Dynamic equilibrium

At equilibrium, there is a balance: reactants and products are both present and the reaction *appears* to have stopped.

- Although there is no apparent change, both forward and reverse processes continue to take place – the equilibrium is *dynamic*.

- The forward reaction proceeds at the **same rate** as the reverse reaction.

> A reversible reaction can be approached from either direction and the terms *reactants* and *products* need to be used with caution.

- The concentrations of reactants and products are constant.

The equilibrium position:

- can be reached from either forward or reverse directions
- can only be achieved in a *closed system* – one in which no materials are being added or removed.

Equilibrium may be approached from either direction

Only in a closed system

Factors affecting equilibrium

AQA	M2	OCR	M2
EDEXCEL	M2	SALTERS	M2
CCEA	M2	WJEC	M1

By opening up a closed equilibrium system, conditions can be changed after which the system can be allowed to reach equilibrium again.

The equilibrium position of the system may be altered by the following changes:

- Changing the concentration of a reactant or product.
- Changing the pressure of a gaseous equilibrium.
- Changing the temperature.

The likely effect on the equilibrium position can be predicted using Le Chatelier's Principle.

> **KEY POINT**
>
> *Le Chatelier's Principle* states that if a system in dynamic equilibrium is subjected to a change, processes will occur to minimise this change.

The effect of concentration changes

A change in the concentration of a reactant will alter the rate of the forward direction. A change in the concentration of a product will alter the rate of the reverse direction. Either change results in a shift in the equilibrium position.

$$N_2(g) + 3H_2(g) \rightleftharpoons 2NH_3(g)$$

Change: **Increase** concentration of a reactant: $N_2(g)$ or $H_2(g)$

Change opposed: The equilibrium system will **decrease** the concentration of the reactant by removing it. This is achieved by a shift in equilibrium to the right, forming more $NH_3(g)$.

The effects of changes in concentration for this equilibrium are shown below. These effects will apply to any system in equilibrium.

> If the concentration of a reactant is increased, or the concentration of the product is decreased, e.g. by removing some of it, the equilibrium is displaced to the right and more product is obtained.

> **KEY POINT**
>
> increase concentration of reactant OR reduce concentration of product
>
> $\longrightarrow$
>
> $$N_2(g) + 3H_2(g) \rightleftharpoons 2NH_3(g)$$
>
> $\longleftarrow$
>
> decrease concentration of reactant OR increase concentration of product

The effect of pressure changes

A change in total pressure may alter the equilibrium position of a system involving gases. **The direction favoured depends upon the total number of gas molecules on each side of the equilibrium.**

$$N_2(g) + 3H_2(g) \rightleftharpoons 2NH_3(g)$$

Change: **Increase** the total pressure.

Change opposed: The equilibrium system will **decrease** the pressure by reducing the total number of moles of gas molecules. This is achieved by a shift in equilibrium to the right:

$$N_2(g) \quad + \quad 3H_2(g) \rightleftharpoons 2NH_3(g)$$
$$1 \text{ mol} \quad\quad\quad 3 \text{ mol} \quad\quad\quad 2 \text{ mol}$$

$$4 \text{ mol} \quad\quad\quad\quad\quad 2 \text{ mol}$$

> Movement of gas molecules causes pressure. The more gas molecules in a volume, the greater the pressure.

> If there are more moles of gaseous reactant than there are moles of gaseous product, an increase in total pressure will displace the reaction to the right.

Increasing the pressure also increases the concentration of any gas present and this **speeds** up the reaction.

The equilibrium position is also changed if:

- at least one of the equilibrium species is a gas
- there are different numbers of gaseous moles of reactants and of products.

The effect of temperature changes

A change in temperature alters the rates of the forward and reverse reactions by different amounts, resulting in a shift in the equilibrium position. **The direction favoured depends upon the sign of the enthalpy change.**

$$N_2(g) + 3H_2(g) \rightleftharpoons 2NH_3(g) \quad\quad \Delta H^\ominus = -92 \text{ kJ mol}^{-1}$$

Change: **Decrease** the temperature.

Change opposed: The equilibrium system will **increase** the temperature by releasing more heat energy. This is achieved by a shift in equilibrium in the exothermic direction, to the right.

> An exothermic process is favoured by low temperatures.
> An endothermic process is favoured by high temperatures.

> An exothermic reaction in one direction is endothermic in the opposite direction.
> The value of the enthalpy changes is the same – the sign is different.

	exothermic process favoured by low temperatures		KEY POINT
$\Delta H^\ominus = +92 \text{ kJ mol}^{-1}$	$N_2(g) + 3H_2(g) \rightleftharpoons 2NH_3(g)$	$\Delta H^\ominus = -92 \text{ kJ mol}^{-1}$	
	endothermic process favoured by high temperatures		

Increasing the temperature *speeds* up the reaction and less time is needed to reach equilibrium. However, the equilibrium position may change to produce a lower equilibrium yield of products (see also pages 113–115).

The effect of a catalyst

There is **no change** in the equilibrium position. However, the catalyst **speeds** up both forward and reverse reactions and equilibrium is reached quicker.

The hydrogencarbonate / carbonate equilibrium

WJEC ▸ M1

Carbon dioxide is an acidic gas that is slightly soluble in water. Dissolved carbon dioxide forms an acidic solution, containing H^+ ions and hydrogencarbonate ions, HCO_3^-.

$$CO_2(aq) + H_2O(l) \rightleftharpoons H^+(aq) + HCO_3^-(aq)$$

Any increase in CO_2 would move this equilibrium to the right, increasing the acidity of the water by increasing the H^+ concentration.

In sea-water, a second natural equilibrium exists between hydrogencarbonate ions, HCO_3^-, and carbonate ions, CO_3^{2-}.

$$HCO_3^-(aq) \rightleftharpoons H^+(aq) + CO_3^{2-}(aq)$$

If the H^+ concentration of sea-water increases, this equilibrium will move to the left to reduce the change.

This self-regulating process maintains the acidity of sea-water at a near constant pH.

Equilibria, rates and industrial processes

AQA ▸ M2   OCR ▸ M2
CCEA ▸ M2

For many industrial processes, the overall yield of a chemical product can vary considerably, depending on the conditions used.

In deciding the conditions to use, industrial chemists need to consider:

All courses study one or more industrial processes.

You should learn outline details of any process on your course.

The principles, however, are common for any similar situation.

- the availability of the starting materials required for a process
- the equilibrium conditions required to ensure a good yield
- the rate of reaction, which should be fast but manageable
- the cost, which should be as low as possible taking into account energy, cost of materials and of the chemical plant
- the safety of workers from hazardous chemicals, pressure, temperature, etc.
- any effects on the environment from waste discharges, toxic fumes, etc.

Industrial chemists compare each of these factors to arrive at compromise conditions. Although the equilibrium yield might be **optimised** by using a very high temperature and pressure, this may prove **impracticable** for reasons of cost and safety.

The importance of compromise in industrial processes

AQA ▸ M2   OCR ▸ M2
CCEA ▸ M2

Four industrial processes are discussed below to illustrate the importance of compromise.

In each of the four processes:

- the right-hand side, with the desired product, has **fewer moles of gas** than the left-hand side (reactants)
- the forward reaction, forming the desired product, is **exothermic**.

The raw materials used should be readily available.

$N_2(g)$ is obtained from the air.

$H_2(g)$ is obtained, together with $CO(g)$, by reacting together natural gas, $CH_4(g)$, and steam, $H_2O(g)$.

$C_2H_4(g)$ is obtained from the cracking of crude oil fractions.

The Haber process for the production of ammonia

$$N_2(g) + 3H_2(g) \rightleftharpoons 2NH_3(g) \qquad \Delta H^\ominus = -92 \text{ kJ mol}^{-1}$$

The contact process during the production of sulfuric acid

$$2SO_2(g) + O_2(g) \rightleftharpoons 2SO_3(g) \qquad \Delta H^\ominus = -197 \text{ kJ mol}^{-1}$$

The industrial production of ethanol by hydration of ethene

$$C_2H_4(g) + H_2O(g) \rightleftharpoons C_2H_5OH(g) \qquad \Delta H^\ominus = -46 \text{ kJ mol}^{-1}$$

The industrial production of methanol from carbon monoxide with hydrogen

$$CO(g) + 2H_2(g) \rightleftharpoons CH_3OH(g) \qquad \Delta H^\ominus = -91 \text{ kJ mol}^{-1}$$

The optimum equilibrium conditions

Optimising the process
These are the ideal conditions to give a maximum equilibrium yield.

Consider le Chatelier's Principle:
- The forward reaction produces fewer moles of gas, favoured by a **high** pressure.
- The forward reaction producing the desired product is exothermic, favoured by a **low** temperature.

The optimum equilibrium conditions for maximum equilibrium yield are:
- **high pressure and low temperature.**

The need for compromise

Compromising
Optimum conditions, reaction rate, feasibility, reality, safety and economics must be considered.

Having arrived at optimum equilibrium conditions, each must be considered to arrive at compromise conditions.

optimum condition	advantages	disadvantages
HIGH PRESSURE	equilibrium yield of desired product is high	energy costs are high – it is expensive to compress gases
	the concentration of gases is high, increasing the rate	there are considerable safety implications of using very high pressures – vessel walls need to be very thick to withstand high pressures, and weaknesses cause danger to workers and potential leakage of chemicals into the environment
LOW TEMPERATURE	equilibrium yield of product is high	the reaction takes place very slowly as few molecules possess the activation energy of the reaction

The use of a catalyst

A catalyst speeds up the rates of both the forward and backward reactions and hence less time is needed for the reaction to reach equilibrium. The increase in reaction rate allows lower temperatures to be used for a realistic reaction rate. The use of lower temperatures has the added bonus of supporting the optimum conditions for this process, increasing the equilibrium yield of the desired product.

Compromise conditions

Exam papers always test the wider aspects of chemistry to society.

Exam answers need to be sensible and related to relevant issues arising from the process itself. It is pointless to use terms such as 'dangerous', 'toxic', etc. unless substantiated with the reasons why the hazard is present.

The compromise conditions used in each process are such that:
- the temperature is sufficiently high to allow the reaction to occur at a realistic rate, but not too high to give a minimal equilibrium yield of the desired product
- a high pressure is used but not too high as to be impracticable
- the reaction rate is increased by using a catalyst. This enables the process to take place at lower temperatures, saving on energy costs.

The compromise conditions used for a process may only yield a small percentage of the desired product. Unreacted reactants are recycled.

Use of catalysts in industry

EDEXCEL M2
OCR M2
WJEC M2

Catalysis is of great economic importance, for example:

iron in the Haber process;

vanadium(V) oxide, V_2O_5, in the contact process;

Ziegler–Natta catalyst in poly(ethene) production;

platinum / palladium / rhodium in catalytic converters.

Industrial preparations often use catalysts, allowing reactions to occur that would not otherwise take place.

Catalysis can reduce production costs and also lead to benefits for the environment.

- Catalysts affect the conditions that are needed, often requiring lower temperatures.
- With lower temperatures, energy demand is reduced. If energy is generated by the combustion of fossil fuels, there will be fewer CO_2 emissions.
- Catalysts may allow different reactions to be used, with better atom economy and reduced waste. (See also page 124.)
- Catalysts are often enzymes, generating very specific products, and operating effectively close to room temperature and pressure. This increases percentage yields and saves on energy. Conventional catalysts are often toxic and their disposal may present environmental problems. Enzymes are usually non-toxic, making for a 'greener' process.

Progress check

1 What will be the result of an increase in pressure on the following equilibria?
   (a) $N_2O_4(g) \rightleftharpoons 2NO_2(g)$
   (b) $CO(g) + 2H_2(g) \rightleftharpoons CH_3OH(g)$
   (c) $H_2(g) + Br_2(g) \rightleftharpoons 2HBr(g)$

2 What will be the result of an increase in temperature on the following equilibria?
   (a) $N_2(g) + O_2(g) \rightleftharpoons 2NO(g)$ $\Delta H^\ominus = +180$ kJ mol^{-1}
   (b) $H_2(g) + Br_2(g) \rightleftharpoons 2HBr(g)$ $\Delta H^\ominus = -9.6$ kJ mol^{-1}
   (c) $2SO_2(g) + O_2(g) \rightleftharpoons 2SO_3(g)$ $\Delta H^\ominus = -197$ kJ mol^{-1}

3 What are the optimum conditions of temperature **and** pressure for a high yield of the products in the equilibria below?
   (a) $CO(g) + 2H_2(g) \rightleftharpoons CH_3OH(g)$ $\Delta H^\ominus = -92$ kJ mol^{-1}
   (b) $PCl_5(g) \rightleftharpoons PCl_3(g) + Cl_2(g)$ $\Delta H^\ominus = +124$ kJ mol^{-1}

3 (a) low temperature and high pressure; (b) high temperature and low pressure.
2 (a) moves to right; (b) moves to left; (c) moves to left.
1 (a) moves to left; (b) moves to right; (c) no change.

Sample question and model answer

1

The diagram below shows the Boltzmann distribution at a temperature, T_1, for a mixture of gases that react with each other. The activation energy for the reaction is labelled **X**.

Notice that the distribution curve starts at the origin but it never quite reaches the x-axis, even at high energies. Remember also that the area under the curve is equal to the total number of molecules.

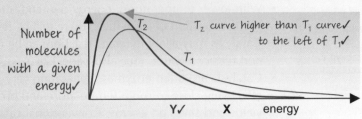

Number of molecules with a given energy ✓

T_2 curve higher than T_1 curve ✓
to the left of T_1 ✓

Y ✓   X   energy

(a) (i) Explain the meaning of the term *activation energy*.

This is the minimum ✓ energy required for a reaction to take place. ✓

(ii) Label the y-axis.

At different temperatures,
the curve changes
the activation energy stays the same.

(iii) Draw a second curve on the diagram above for the same mixture at a lower temperature. Label the second curve T_2.

Note the difference

(iv) Explain how a lower temperature affects the rate of this reaction.

Reduced temperature slows down the reaction rate. ✓
The overall kinetic energy of the molecules is reduced ✓ and
fewer molecules now exceed the activation energy. ✓   [8]

In the presence of a catalyst,
the activation energy changes
the curve stays the same.

(b) A catalyst is added to the mixture of gases.

(i) On the diagram above, label a possible activation energy **Y** for the catalysed reaction.

(ii) Explain, in terms of activation energy and the Boltzmann distribution, how a catalyst affects the rate of this reaction.

The catalyst speeds up the reaction rate. ✓ A catalyst allows the
reaction to proceed via a different route ✓ with a lower activation
energy. This means that a larger proportion of molecules now exceeds
the activation energy. ✓   [4]

(c) The equation for the industrial manufacture of ethanol, C_2H_5OH, from ethene, C_2H_4, is shown below.

$$C_2H_4(g) + H_2O(g) \rightleftharpoons C_2H_5OH(g) \quad \Delta H = -46 \text{ kJ mol}^{-1}$$

Always pay attention to the words that examiners use.

In parts (i) and (ii), the examiner has used 'Explain'. You need to justify your answer.

In part (iii), the examiner has used 'State'. You only need to 'state' your answer. You waste time here if you 'explain'.

The industrial conditions used are a high temperature and high pressure.

(i) Explain why the reaction is carried out at a high pressure.
There is a greater equilibrium yield. ✓ The increase in pressure
is relieved by reducing the number of gas molecules. ✓ The
equilibrium shifts in favour to the right because there are fewer
gas molecules to the right. ✓

(ii) Explain why pressures higher than 1000 atmospheres are not used.
High pressures are very costly to generate in terms of energy. ✓

(iii) State **one** advantage and **one** disadvantage of carrying out the reaction at high temperature.
An advantage is that there is a greater reaction rate. ✓
A disadvantage is that the equilibrium yield is reduced. ✓   [6]

[Total: 18]

Practice examination questions

1 (a) Write a chemical equation, including state symbols, for the reaction that is used to define the enthalpy change of formation of one mole of sodium carbonate, $Na_2CO_3(s)$. [2]

(b) State the standard conditions used for a standard enthalpy change of formation, $\Delta H^{\ominus}_f$. [1]

(c) Use the standard enthalpy changes of formation given below to calculate a value for the standard enthalpy change for the following reaction:
$$Na_2CO_3.10H_2O(s) \longrightarrow Na_2CO_3(s) + 10H_2O(l)$$

compound	$Na_2CO_3.10H_2O(s)$	$Na_2CO_3(s)$	$H_2O(l)$
$\Delta H^{\ominus}_f$ / kJ mol^{-1}	−4081	−1131	−286

[3]

[Total: 6]

2 Propan-1-ol, $CH_3CH_2CH_2OH$, reacts with oxygen in a combustion reaction.

(a) (i) Define the term *standard enthalpy change of combustion*.
(ii) State the temperature that is conventionally chosen for standard enthalpy changes. [4]

(b) (i) Write a balanced equation for the combustion of propan-1-ol.
(ii) Calculate the standard enthalpy change of combustion of propan-1-ol using the following data.

compound	$\Delta H^{\ominus}_f$ /kJ mol^{-1}
$CH_3CH_2CH_2OH(l)$	−303
$CO_2(g)$	−394
$H_2O(l)$	−286

[4]

[Total: 8]

3 (a) Define the term *standard enthalpy change of formation* ($\Delta H^{\ominus}_f$). [3]

(b) Give the equation for the change that represents the standard enthalpy change of formation of methane. [2]

(c) Use the following data to calculate a value for the standard enthalpy change of formation of methane.

compound	$\Delta H^{\ominus}_c$ /kJ mol^{-1}
C(s)	−394
$H_2(g)$	−242
$CH_4(g)$	−802

[3]

(d) Use the data below to calculate average bond enthalpy values for the C–H and the C–C bonds.

$$CH_4(g) \longrightarrow C(g) + 4H(g) \qquad \Delta H = 1648 \text{ kJ mol}^{-1}$$
$$C_2H_6(g) \longrightarrow 2C(g) + 6H(g) \qquad \Delta H = 2820 \text{ kJ mol}^{-1} \qquad [3]$$

[Total: 11]

Practice examination questions

4 (a) The diagram below shows a Boltzmann distribution for the mixture of gases at a temperature, T_1.

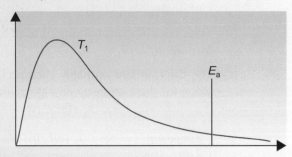

  (i) Label the axes and add a second curve for the Boltzmann distribution of this mixture at a higher temperature, T_2.

  (ii) On the diagram, add a label: 'E_a catalysed' for a possible activation energy of the reaction when catalysed. [5]

(b) The elimination of steam from ethanol is an endothermic reaction:

$$C_2H_5OH(g) \longrightarrow C_2H_4(g) + H_2O(g) \qquad \Delta H = +46 \text{ kJ mol}^{-1}$$

  (i) Sketch the reaction profile diagram for this reaction. On your diagram, label clearly the enthalpy change for the reaction, ΔH, and the activation energy, E_a.

  (ii) Add to your diagram a labelled reaction profile diagram for this reaction when catalysed.

  (iii) Explain why a catalyst increases the rate of a chemical reaction.

[7]

[Total: 12]

5 When hydrogen and carbon dioxide react, a dynamic homogenous equilibrium is set up as shown below:

$$H_2(g) + CO_2(g) \rightleftharpoons H_2O(g) + CO(g)$$

(a) (i) Why is this reaction a *homogeneous equilibrium*?

  (ii) State **three** features of a *dynamic equilibrium*. [5]

(b) For this equilibrium, explain the effect of an increase in pressure on:

  (i) the equilibrium position

  (ii) the reaction rate. [4]

(c) The equilibrium yield of steam and carbon monoxide increases as the temperature is increased. Determine whether the forward reaction in the equilibrium above is exothermic or endothermic. Explain your answer. [2]

[Total 11]

Chapter 4

Organic chemistry, analysis and the environment

The following topics are covered in this chapter:

- *Basic concepts*
- *Hydrocarbons from oil*
- *Alkanes*
- *Alkenes*

- *Alcohols*
- *Halogenoalkanes (Haloalkanes)*
- *Analysis*
- *Chemistry in the environment*

4.1 Basic concepts

After studying this section you should be able to:

- *understand the different types of formula used for organic compounds*
- *recognise types of hydrocarbon*
- *recognise common functional groups*
- *understand what is meant by structural isomerism*
- *apply rules for naming simple organic compounds*
- *understand the difference between homolytic and heterolytic fission*
- *calculate percentage yields and atom economies*

LEARNING SUMMARY

Types of formula

AQA	M1	OCR	M2
EDEXCEL	M1	SALTERS	M1
CCEA	M2	WJEC	M2

In organic chemistry, there are many ways of representing a formula.

For the compound butane, with 4 carbon atoms and 10 hydrogen atoms:

the *empirical* formula is:	C_2H_5	The simplest, whole-number ratio of elements in a compound.
the *molecular* formula is:	C_4H_{10}	The *actual* number of atoms of each element in a molecule.
the *structural* formula is:	$CH_3CH_2CH_2CH_3$	The minimal detail for an unambiguous structure.
the *displayed* formula is:		The relative placing of atoms and the bonds between them.
the *skeletal* formula	∧∨	The carbon skeleton and functional groups only.

Carbon chains

AQA	M1	OCR	M2
EDEXCEL	M1	SALTERS	M1
CCEA	M2	WJEC	M2

Hydrocarbons

Hydrocarbons are compounds of carbon and hydrogen **only**.

- A saturated hydrocarbon has single bonds only.
- An unsaturated hydrocarbon contains a multiple carbon carbon bond.

saturated – single bonds only

unsaturated – contains a double bond

Alkanes

Carbon atoms can bond with other carbon atoms to form an enormous range of compounds with different carbon-chain lengths. The simplest organic compounds are a family of saturated hydrocarbons called the alkanes, shown below.

meth- C
eth- C_2
prop- C_3
but- C_4
pent- C_5
hex- C_6
hept- C_7
oct- C_8
non- C_9
dec- C_{10}

Number of carbons	Name	Molecular formula	Structural formula
1	methane	CH_4	CH_4
2	ethane	C_2H_6	CH_3CH_3
3	propane	C_3H_8	$CH_3CH_2CH_3$
4	butane	C_4H_{10}	$CH_3CH_2CH_2CH_3$
5	pentane	C_5H_{12}	$CH_3CH_2CH_2CH_2CH_3$
6	hexane	C_6H_{14}	$CH_3CH_2CH_2CH_2CH_2CH_3$

Note the following points.

- The name of the alkane ends with –*ane*.
- The prefixes (*meth-*, *eth-*, …) are used to represent the number of carbon atoms. You will need to use these many times in organic chemistry and they must be learnt.

General formula

Alkanes, C_nH_{2n+2} are saturated.

- The **general formula** is the simplest algebraic representation for any member in a series of organic compounds.
- The general formula for any alkane is C_nH_{2n+2}, where n = the number of carbon atoms.

Homologous series

Members in a homologous series react similarly.

By studying the reactions of one member of the series, you know how all members in the series are likely to react.

The alkanes are an example of a **homologous series** with the following features.

- Each successive member differs by $-CH_2-$. You can see this by comparing the formula of each alkane in the table above.
- Each member in a homologous series has the same general formula. The alkanes have the general formula: C_nH_{2n+2}.
- All members in a homologous series have the same functional group and similar chemical reactions.

Note that physical properties, such as boiling point and density, do gradually change as the length of the carbon chain increases (see also page 127).

Functional groups

AQA	M1	OCR	M2
EDEXCEL	M1	SALTERS	M1
CCEA	M2	WJEC	M2

Saturated hydrocarbon chains are comparatively unreactive. The reactivity is increased by the presence of a functional group – the reactive part of a carbon compound.

Unreactive carbon chain *alkyl group* C_nH_{2n+1} — Functional group The part responsible for the reactions

Common functional groups

It is essential that you can instantly identify a functional group within a molecule. You can then apply the relevant chemistry.

If you continue to study Chemistry to A Level, you will meet more functional groups.

The main functional groups for AS Chemistry are shown below.

name	functional group	structural formula	general formula	prefix or suffix (for naming)
alkane	C–H	$CH_3CH_2CH_3$ *propane*	C_nH_{2n+2}	-ane
alkene	$\backslash C{=}C /$	$CH_3CH{=}CH_2$ *propene*	C_nH_{2n}	-ene
halogenoalkane	C–X (X=halogen)	CH_3CH_2Br *bromoethane*	$C_nH_{2n+1}Br$	bromo-
alcohol	C–OH	CH_3CH_2OH *ethanol*	$C_nH_{2n+1}OH$	-ol
aldehyde	$-C{\nwarrow}^O_H$	CH_3CHO *ethanal*	$C_nH_{2n}O$	-al
ketone	$R-\overset{O}{\overset{\|}{C}}-R$	$CH_3COC_2H_5$ *butanone*	$C_nH_{2n}O$	-one
carboxylic acid	$-C{\nwarrow}^O_{OH}$	CH_3COOH *ethanoic acid*	$C_nH_{2n+1}COOH$	-oic acid

- When studying reactions of a homologous series, the alkyl group is relatively unimportant – it is the functional group that reacts.
- The alkyl group, C_nH_{2n+1} is often represented simply as R–. This shifts the emphasis in the formula towards the reactive part of the molecule, the functional group.

The alcohols, shown below, is a homologous series with the –OH functional group.

- The alcohols have the general formula:
  $C_nH_{2n+1}OH$
- *Any* alcohol can be represented as:
  R–OH

The alcohols

CH_3OH
C_2H_5OH
C_3H_7OH

Same functional group, OH

Structural isomerism

AQA	M1	OCR	M2
EDEXCEL	M1	SALTERS	M1
CCEA	M2	WJEC	M2

In addition to forming long chains, the atoms making up a molecular formula are often arranged differently, forming **isomers**.

Structural isomerism

Structural isomers are molecules with the same molecular formula but with different structural formulae (structural arrangements of atoms).

The two structural isomers of C_4H_{10} are shown below:

See also *E/Z isomerism*, page 129.

```
   H   H   H   H              H   CH3  H
   |   |   |   |              |    |   |
H—C — C — C — C—H         H—C —  C —  C—H
   |   |   |   |              |    |   |
   H   H   H   H              H    H   H
      butane                   methylpropane
```

Butane and methylpropane are chain isomers of C_4H_{10}. Chain isomerism is a type of structural isomerism in which the carbon skeleton is different.

The number of different structural isomers possible from a molecular formula increases dramatically as the carbon-chain length increases. This is shown in the table below.

molecular formula	number of isomers
C_5H_{12}	3
C_6H_{14}	5
C_7H_{16}	9
C_8H_{18}	18
C_9H_{20}	35
$C_{10}H_{22}$	75
$C_{15}H_{32}$	4,347
$C_{20}H_{42}$	366,319

Look at the different names given to the two isomers.

- The branched isomer, methylpropane, is treated as a side chain attached to a straight-chain alkane.
- Side chains, such as the methyl group $CH_3–$, are called *alkyl* groups with the general formula of C_nH_{2n+1}.
- An alkyl group can be regarded as an alkane with one hydrogen atom removed (to allow another group to be attached).

Notice how the alkyl groups are named and how their formulae are linked to the parent alkane.

You must learn these.

The naming of organic compounds is discussed in more detail below.

Alkanes and alkyl groups

Number of carbons	Alkane C_nH_{2n+2}		Alkyl group C_nH_{2n+1}	
	Formula	Name	Formula	Name
1	CH_4	methane	$CH_3–$	methyl
2	C_2H_6	ethane	$C_2H_5–$	ethyl
3	C_3H_8	propane	$C_3H_7–$	propyl

In position isomerism, a functional group can be at different positions on the chain. For example, there are two position isomers with the molecular formula $C_4H_{10}O$ that have the alcohol, –OH, functional group:

- butan-1-ol with the –OH functional group at the end of the carbon chain
- butan-2-ol with the –OH functional group one carbon in from the end of the carbon chain.

butan-1-ol

butan-2-ol

Naming of organic compounds

AQA	M1	OCR	M2
EDEXCEL	M1	SALTERS	M1
CCEA	M2	WJEC	M2

With so many isomers possible, it is important that each has an individual name. The examples below show the steps needed to name an organic compound.

Rule 1

- The name is based upon the longest carbon chain on an **alkane**.

butane

Rule 2

- Any functional groups and alkyl groups are identified.
- These are then added to the name as a prefix (e.g. chloro-) or suffix (e.g. -ol) (see page 121).

methylbutanol

Rule 3

- If there is more than one possible isomer then the carbon atoms are labelled with numbers. Numbering starts from the end giving the lowest possible numbers for any functional groups and side chains. In this example, numbering from the left would give the incorrect name of 3-methylbutan-4-ol.

2-methylbutan-1-ol

Rule 4

- If there is more than one alkyl or functional group, they are placed in alphabetical order. This rule is not needed for the example above but it does need to be applied to name the compound to the right.

3-ethyl-2-methylpentane

Types of bond fission

AQA	M2	OCR	M2
EDEXCEL	M2	SALTERS	M2
CCEA	M2	WJEC	M2

The breaking of a covalent bond is called **bond fission**. Reactions of organic compounds involve bond fission followed by the formation of new bonds. Two types of bond fission are possible, **homolytic** fission and **heterolytic** fission.

Homolytic fission

> These are very important principles, used throughout organic chemistry.

In homolytic fission, bond-breaking produces two species of the same (*homo-*) type:

$$A{:}B \longrightarrow A{\cdot} + {\cdot}B$$
radicals

> A radical is a species with an unpaired electron (see page 128).

- In the example above, a covalent bond breaks so that one of the bonding electrons goes to each of **A** and **B**.
- Homolytic fission forms **two radicals**.

Heterolytic fission

> Homolytic fission
> $\longrightarrow$ radicals
> Heterolytic fission
> $\longrightarrow$ ions

In heterolytic fission, bond-breaking produces two species of different (*hetero-*) type:

$$A{:}B \longrightarrow A{:}^- + B^+ \qquad \text{or} \qquad A{:}B \longrightarrow A^+ + {:}B^-$$
ions $\qquad\qquad\qquad\qquad\qquad$ *ions*

- In the example above, a covalent bond breaks so that both the bonding electrons go to either **A** or **B**.
- Heterolytic fission forms **oppositely-charged ions**.

Percentage yields

AQA	M1	OCR	M2
EDEXCEL	M1	WJEC	M1
CCEA	M2		

Organic reactions typically produce a lower yield than expected from the balanced equation. The yield is usually expressed as a percentage yield:

> **KEY POINT**
> $$\text{percentage yield} = \frac{\text{actual yield}}{\text{theoretical yield}} \times 100\%$$

Example

> See also page 39.

5.2 g of 1-bromobutane is formed by reacting 5.0 g of butan-1-ol with sodium bromide and concentrated sulfuric acid. Find the percentage yield of 1-bromobutane.

equation	$C_4H_9OH + NaBr + H_2SO_4 \longrightarrow C_4H_9Br + H_2O + NaHSO_4$
moles	1 mol $\longrightarrow$ 1 mol

> Factors that can contribute to a low yield:
> Organic reactions often do not go to completion.
> An organic compound often reacts to produce a mixture of products.
> The purification stages result in loss of some of the desired product.

reacting masses	74.0 g	$\longrightarrow$	136.9 g
	1 g	$\longrightarrow$	$\dfrac{136.9}{74.0}$ g
	5.0 g	$\longrightarrow$	$5 \times \dfrac{136.9}{74.0}$ g

$\therefore$ 5.0 g of C_4H_9OH produces a theoretical yield of 9.3 g C_4H_9Br.

Mass of C_4H_9Br that forms = 5.2 g.

$$\text{percentage yield} = \frac{\text{actual yield}}{\text{theoretical yield}} \times 100\% = \frac{5.2}{9.3} \times 100 = 56\%$$

Atom economy

AQA	M1	OCR	M2
EDEXCEL	M1	SALTERS	M2
CCEA	M2	WJEC	M1

We are all becoming far more aware of our environment and the need to conserve resources and produce less waste. Atom economy is used to assess the efficiency of a reaction in terms of atoms:

> **KEY POINT**
>
> $$\text{atom economy} = \frac{\text{molecular mass of the desired product}}{\text{sum of molecular masses of all products}} \times 100$$

Atom economy is calculated from the balanced equation for the reaction.

Example

> To improve the atom economy for C_4H_9Br production, a chemical company must either find uses for the other products or use another method for C_4H_9Br production.

equation: $C_4H_9OH + NaBr + H_2SO_4 \longrightarrow C_4H_9Br + H_2O + NaHSO_4$

molecular mass of desired product:

$$C_4H_9Br = 136.9$$

sum of molecular masses of all products:

$$C_4H_9Br + H_2O + NaHSO_4 = 136.9 + 18.0 + 120.1 = 275.0$$

$$\text{atom economy} = \frac{136.9}{275.0} \times 100 = 49.8\%$$

The consequence is that, even if this reaction could be carried out with a 100% percentage yield, converting **all** of the organic starting material, C_4H_9OH, into the organic product, C_4H_9Br, more than half the mass of products is **waste**.

Atom economy and type of reaction

> An addition reaction produces a single product: all atoms are used and the reaction has an atom economy of 100%.

Addition reactions are the most atom efficient, giving a single product and an atom economy of 100%.

addition: $CH_2=CH_2 + H_2O \longrightarrow C_2H_5OH$

> Substitution and elimination always produce by-products.
>
> These are wasted atoms, unless the by-products can be used.

Substitution and elimination reactions produce by-products and will always have less atom economy unless some use is found for by-products.

substitution: $CH_3CH_2CH_2Br + NaOH \longrightarrow CH_3CH_2CH_2OH + NaBr$

elimination: $CH_3CH_2CH_2Br \longrightarrow CH_3CH=CH_2 + HBr$

Progress check

1 C_6H_{14} has 5 structural isomers. Show the structural and displayed formula for each of these isomers and name them.

2 8.52 g of 1-chloropentane was reacted with aqueous sodium hydroxide, producing 4.75 g of pentan-1-ol.
$CH_3CH_2CH_2CH_2CH_2Cl + NaOH \longrightarrow CH_3CH_2CH_2CH_2CH_2OH + NaCl$
Find the percentage yield of 1-bromobutane and the atom economy of the reaction.

2 Percentage yield: 67.5 atom economy: 82.6%.

124

4.2 Hydrocarbons from oil

After studying this section you should be able to:

- *understand how hydrocarbon fractions are separated from crude oil*
- *describe how low demand fractions are processed into hydrocarbons of higher demand*

LEARNING SUMMARY

Useful products from crude oil

AQA	M1	OCR	M2
EDEXCEL	M1	SALTERS	M1
CCEA	M2	WJEC	M2

Crude oil is a fossil fuel, formed from the decay of sea creatures over millions of years. It is a complex mixture of hydrocarbons, containing mainly alkanes. Crude oil cannot be used directly but it is the source of many useful products. The first stages in the processing of crude oil are described below.

Fractional distillation

Crude oil is heated and passed into a fractionating tower, separating the complex mixture into *fractions*. Note that the fractions are not pure and contain mixtures of hydrocarbons within a range of boiling points. The fractionating tower, shown below, shows the temperature gradient which separates the fractions according to their boiling points (see also page 127).

Fractions with the lowest boiling points are collected at the top of the tower.

On descending the tower, the temperature rises and the boiling points increase.

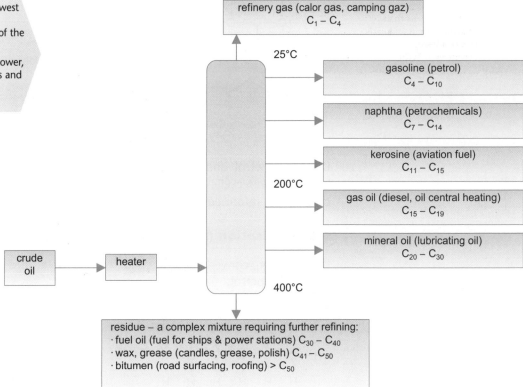

refinery gas (calor gas, camping gaz)
$C_1 - C_4$

25°C

gasoline (petrol)
$C_4 - C_{10}$

naphtha (petrochemicals)
$C_7 - C_{14}$

kerosine (aviation fuel)
$C_{11} - C_{15}$

200°C

gas oil (diesel, oil central heating)
$C_{15} - C_{19}$

mineral oil (lubricating oil)
$C_{20} - C_{30}$

crude oil → heater →

400°C

residue – a complex mixture requiring further refining:
· fuel oil (fuel for ships & power stations) $C_{30} - C_{40}$
· wax, grease (candles, grease, polish) $C_{41} - C_{50}$
· bitumen (road surfacing, roofing) $> C_{50}$

Cracking

Cracking is the starting point for the manufacture of many organics.

Cracking is the breaking down of an unsaturated hydrocarbon into smaller hydrocarbons. Cracking breaks C–C bonds in alkanes.

The purpose of cracking is to produce high demand hydrocarbons:

- short-chain alkanes for use in petrol
- alkenes, as a feedstock for a wide range of organic chemicals, including polymers (see also page 132).

125

Thermal cracking heats the *naphtha fraction* with steam at a high temperature (about 800°C) and high pressure. This forms a mixture of straight-chain alkanes and alkenes (mainly ethene) with a small proportion of branched and cyclic hydrocarbons. Some hydrogen is also produced.

The following equations summarise the overall process of cracking.

> **Thermal cracking** breaks C–C bonds by homolytic fission forming radicals.
>
> **Catalytic cracking** breaks C–C bonds by heterolytic fission forming ions.

$$C_{10}H_{22} \longrightarrow C_8H_{18} + C_2H_4$$
$$C_{10}H_{22} \longrightarrow C_6H_{14} + 2C_2H_4$$
$$\text{alkane} \longrightarrow \text{alkane} + \text{alkene}$$

> There are many more equations possible but all must produce both an alkane and an alkene.

Catalytic cracking processes heavy long-chain fractions obtained in larger quantities than required. A mixture of the vapourised fraction and a zeolite catalyst are reacted at about 450°C using a slight pressure only. The process forms a higher proportion of branched and cyclic hydrocarbons than thermal cracking (see also reforming and isomerisation below). These are used for petrol.

Reforming

> In a car engine, petrol has the tendency to auto-ignite before the spark, causing 'knocking'. Resistance to auto-ignition is measured as the **octane number** of the fuel.
>
> Straight-chain hydrocarbons have lower octane numbers than branched-chain and cyclic hydrocarbons.

Reforming uses heat and pressure to convert unbranched fractions into cycloalkanes (e.g. cyclohexane) and arenes (e.g. benzene). These products are used in petrol and as a feedstock for a wide range of organic chemicals including many pharmaceuticals and dyes.

Examples

cyclohexane

benzene

Unleaded petrol and lead-replacement petrol require a higher proportion of branched and cyclic hydrocarbons for effective combustion. This has increased demand for reformed alkanes.

Isomerisation (not Edexcel)

> Chemists are improving fuels by use of oxygenates such as methanol and ethanol. Hydrogen is also being developed as a fuel to replace petrol in the future.

Isomerisation converts unbranched hydrocarbons into branched hydrocarbons, needed for petrol.

Example

Progress check

1 Write down the formula of an alkane present in each of the main fractions obtained from crude oil.
2 Construct equations for two possible reactions taking place during the cracking of $C_{14}H_{30}$.

1 Any alkane with a formula C_nH_{2n+2} (see page 125) to match each fraction.
2 Any 2 equations ⟶ alkane + alkene, e.g. $C_{14}H_{30} \longrightarrow C_8H_{18} + C_6H_{12}$

4.3 Alkanes

After studying this section you should be able to:

- *understand why different alkanes have different melting and boiling points*
- *understand that combustion of alkanes provides useful energy*
- *describe the substitution reactions of alkanes with halogens*

LEARNING SUMMARY

Alkanes
General formula: C_nH_{2n+2}

Alkanes are generally unreactive. Alkanes contain only C–H and C–C bonds, which are relatively strong and difficult to break. The similar electronegativities of carbon and hydrogen give molecules which are non-polar. Alkanes are the typical 'oils' used in many non-polar solvents and they do not mix with water.

Melting points and boiling points of alkanes

EDEXCEL	M2	OCR	M2
CCEA	M2	SALTERS	M2
		WJEC	M2

Boiling points increase with increasing chain length because of the increasing strength of the intermolecular forces. This allows the alkane mixture in crude oil to be separated by fractional distillation.

> The same principle can be applied to any homologous series.

Increasing chain length

m.pt and b.pt increase

more surface of contact between molecules

greater van der Waals' forces between molecules

C_3H_8 C_5H_{12}

For hydrocarbons of similar molecular mass, branched alkanes have lower boiling points because there are fewer interactions possible between molecules:

> The surface contact between molecules is important because it allows interaction between electrons. Remember that van der Waals' forces result from fluctuations in electron density (see page 64).

Isomers of C_4H_{10}

m.pt and b.pt increase

more surface of contact between molecules

greater van der Waals' forces between molecules

Combustion of alkanes

AQA	M1	OCR	M2
EDEXCEL	M1	SALTERS	M1
CCEA	M2	WJEC	M2

Combustion with oxygen is the most important reaction of alkanes giving their immediate use as fuels:

natural gas: $CH_4 + 2O_2 \longrightarrow CO_2 + 2H_2O$
petrol: $C_8H_{18} + 12\frac{1}{2}O_2 \longrightarrow 8CO_2 + 9H_2O$

The burning of fossil fuels such as alkanes provides society with many environmental problems, some of which are listed below.

> Although generally unreactive, alkanes do react with oxygen and the halogens.

- Petroleum fractions contain sulfur impurities. Unless removed before combustion, sulfur oxides are formed from which acid rain (H_2SO_3 and H_2SO_4) is produced. The base, calcium oxide, is used to remove sulfur oxides from flue gases in power stations:

> Remember that reactions with oxygen produce oxides.

$$CaO(s) + SO_2(g) \longrightarrow CaSO_3(s)$$

The $CaSO_3$ is oxidised to gypsum, $CaSO_4 \bullet 2H_2O$, which is used to make plaster.

- Toxic gases such as carbon monoxide, nitrogen oxides and unburnt hydrocarbons are present in car emissions. The development of catalytic converters has helped to remove these polluting gases (see also page 109).

> Carbon dioxide, methane and water vapour are the main greenhouse gases that contribute to global warming (see page 147).

- Excessive combustion of fossil fuels increases carbon dioxide emissions contributing to global warming.

127

Substitution of alkanes with halogens

AQA	M2	OCR	M2
EDEXCEL	M1	SALTERS	M2
CCEA	M2	WJEC	M2

> **KEY POINT**
> A *substitution reaction* involves the swapping over of one species for another.

Alkanes are substituted by halogens, such as chlorine and bromine.

$$CH_4 + Cl_2 \longrightarrow CH_3Cl + HCl$$

This reaction only takes place in the presence of ultraviolet radiation, which produces highly reactive radicals.

> You may have used the term 'free radical'. Radical and free radical mean the same thing.

> **KEY POINT**
> A radical, such as Cl•, CH_3•,
> • is a highly reactive species with an unpaired electron
> • reacts by pairing of an unpaired electron
> • is often involved in chain reactions.

Mechanism of radical substitution

Initiation

In this stage, the reaction starts.

> Make sure that you learn this mechanism – these are easy marks in exams if you do!

Ultraviolet radiation provides energy to break Cl–Cl bonds homolytically producing chlorine radicals, Cl•

$$Cl_2 \longrightarrow 2Cl•$$

Propagation

In this stage, the reaction products are made.

Radicals are recycled in a chain reaction

> The chlorine radicals catalyse the reaction.
> See also the breakdown of ozone, pages 109 and 149.

$$CH_4 + Cl• \longrightarrow CH_3• + HCl$$
$$CH_3• + Cl_2 \longrightarrow CH_3Cl + Cl•$$

Termination

In this stage, radicals react together and are removed from the reaction mixture.

$$Cl• + Cl• \longrightarrow Cl_2$$
$$CH_3• + Cl• \longrightarrow CH_3Cl$$
$$CH_3• + CH_3• \longrightarrow CH_3CH_3$$

Impure products

> This is answered poorly in exams.

Further substitution of the reaction products is possible, producing a mixture of products:

$$CH_3Cl \xrightarrow{Cl•} CH_2Cl_2 \xrightarrow{Cl•} CHCl_3 \xrightarrow{Cl•} CCl_4$$

Progress check

1 Write an equation for the complete combustion of butane.
2 Butane reacts with chlorine in a substitution reaction.
  (a) What conditions are essential for this reaction? Explain your answer.
  (b) Write an equation for this reaction.
  (c) Write the structural formula of 2 possible monosubstituted products.
  (d) What is the formula for the final product obtained from complete substitution?

(d) C_4Cl_{10}
(c) Any 2 isomers of C_4H_9Cl, e.g. $CH_3CH_2CH_2CH_2Cl$ and $CH_3CH_2CHClCH_3$
(b) $C_4H_{10} + Cl_2 \longrightarrow C_4H_9Cl + HCl$
2 (a) UV, to generate radicals from Cl_2
1 $C_4H_{10}(g) + 6\frac{1}{2}O_2(g) \longrightarrow 4CO_2(g) + 5H_2O(l)$

4.4 Alkenes

After studying this section you should be able to:

- recognise the nature of *E/Z* isomerism in alkenes
- understand that the double bond in alkenes takes part in addition reactions with electrophiles
- understand the use of alkenes in the production of organic compounds
- understand what is meant by addition polymerisation

LEARNING SUMMARY

Alkenes

General formula: C_nH_{2n}

AQA	M2	OCR	M2
EDEXCEL	M1	SALTERS	M2
CCEA	M2	WJEC	M2

Alkenes are unsaturated compounds with a C=C double bond. The high electron density of the double bond makes alkenes more reactive than alkanes.

Naming of alkenes

Position isomerism

A type of structural isomerism in which the functional group is in a different position on the carbon skeleton.

The position of the double bond is indicated using one number only, as shown in the example below for two isomers of C_4H_8.

Only one number is needed. In but-1-ene, a double bond starting at carbon-1 must finish at carbon-2

but-1-ene but-2-ene

What is a double bond?

The coverage here is greatly simplified but sufficient for AS Chemistry.

In saturated compounds, the C–H and C–C single bonds are σ-bonds. A σ-bond is on the C–H or C–C axis, formed by overlap of s- and p- atomic orbitals.

In unsaturated compounds, the C=C double bond comprises a σ- and a π-bond.

σ- and π-bonds are *molecular orbitals* – these bond together the atoms in a molecule.

A π-bond is above and below the C–C axis, formed by overlap of atomic p-orbitals. The π-bond introduces an area of high electron density in the molecule, shown in the diagram of ethene below.

overlap forms a

σ-bond

π-bond

p-orbitals

The C=C bond
planar molecule
3 σ-electron pairs around each carbon atom.
Bond-angle is approximately 120°

E/Z isomerism

AQA	M2	OCR	M2
EDEXCEL	M1	SALTERS	M2
CCEA	M2	WJEC	M2

Stereoisomers are molecules with the same structural formula but with a different arrangement in space.

KEY POINT

See also structural isomerism, page 121.

E/Z **isomerism** is a type of stereoisomerism that is present in some unsaturated hydrocarbons with a C=C bond.

Single C–C bonds can rotate. The C=C double bond, however, restricts rotation and prevents groups from moving from one side of the double bond to the other. This can give rise to *E/Z* isomerism.

For *E/Z* isomerism, there must be:

No rotation possible about a C=C bond.

- a C=C double bond
- two **different** groups attached to **each** carbon end of the double bond.

If two of the groups are the same, this is also referred to as *cis-trans* isomerism:

- the *E* isomer (*trans* isomer) has the same groups on opposite sides
- the *Z* isomer (*cis* isomer) has the same groups on the same side.

The *E* and *Z* isomers of but-2-ene are shown below:

E-but-2-ene *trans*   *Z*-but-2-ene *cis*

Cis-trans isomerism is limited to unsaturated compounds in which two of the different groups are the same (as in the example above). The *E/Z* naming system can be applied to compounds with more than 2 different groups attached to either end of the C=C double bond.

The *E/Z* isomers of 3-methylhept-3-ene are shown below:

It is impossible to label these as *cis* and *trans* isomers. The rules for labelling each stereoisomer in this example as *E* and *Z* go beyond the demands of AS Level Chemistry.

Addition reactions of alkenes

AQA	M2	OCR	M2	
EDEXCEL	M1	SALTERS	M2	
CCEA	M2	WJEC	M2	

An addition reaction of an alkene involves the opening of the double bond with the formation of a saturated addition product.

> **KEY POINT**
>
> An *addition reaction* involves two species adding together to make one.
> Addition usually converts an **unsaturated** compound into a **saturated** compound.

This is made possible by the high electron density of the π-bond which attracts electrophiles.

> **KEY POINT**
>
> An *electrophile*, such as Br_2, HBr, H_2SO_4, NO_2^+:
> - is an electron-deficient species
> - 'attacks' an electron-rich carbon atom or double bond by **accepting a pair of electrons**.

This type of reaction is called **electrophilic addition**.

Addition of bromine

The addition reaction between an alkene and bromine, in which the orange colour of bromine is decolourised, is used as a test for unsaturation.

Mechanism of electrophilic addition

In mechanisms, *a curly arrow* is used to show the movement of a **pair of electrons**.
The movement of the electron pair involves either:
• the formation of a new covalent bond or
• the heterolytic fission of a covalent bond (see page 123).
The *curly arrow* always goes **from** an electron pair (or double bond).

Remember that a curly arrow shows the movement of an electron pair.

Addition using bromine water (Edexcel only)

Most specifications simplify the reaction of alkenes with bromine water as the addition of bromine. However, Edexcel discusses this reaction in more detail in terms of addition by Br (from Br_2) and OH (from H_2O) across the double bond:

$$H_2C{=}CH_2 + Br_2 + H_2O \longrightarrow BrCH_2CH_2OH + HBr$$

Addition using acidified manganate(VII) (Edexcel only)

Another useful test for unsaturation is the use of cold, dilute acidified manganate(VII). This oxidises the double bond of the alkene with the formation of a diol. The purple colour of manganate(VII) is decolourised.

Note that [O] is used in equations to represent an oxidising agent. Many candidates in exams get this reaction wrong by incorrectly adding Mn. $KMnO_4$ is used as a source of manganate (VII) ions.

Reduction with hydrogen

Reduction of an organic compound involves loss of oxygen OR gain of hydrogen (accompanied by a gain of electrons).

Alkenes are reduced when treated with hydrogen gas in the presence of a nickel catalyst at 150°C. The process of adding hydrogen across a double bond is sometimes referred to as **hydrogenation**.

Hydrogenation of C=C is used in the production of solid margarine from unsaturated liquid vegetable oils. Saturated fats are solid and the amount of hydrogenation can be controlled to give the correct texture of margarine.

Further addition reactions of ethene

Addition reactions of alkenes are useful in organic synthesis. Some further addition reactions of ethene are shown in the flowchart below.

For more details:

polymerisation, see below;

preparation of alcohols, see also page 135.

Although ethanol is made industrially using steam and phosphoric acid H_3PO_4, an alternative synthesis uses:

- addition with concentrated sulfuric acid at room temperature

$$C_2H_4 + H_2SO_4 \longrightarrow CH_3CH_2OSO_3H$$

- followed by hydrolysis

$$CH_3CH_2OSO_3H + H_2O \longrightarrow CH_3CH_2OH + H_2SO_4$$

Addition reactions of unsymmetrical alkenes

These structures are shown below.

In the addition of HBr to propene $CH_3CH=CH_2$, two products are possible:

- the secondary bromoalkane $CH_3CHBrCH_3$ as the major product
- the primary bromoalkane $CH_3CH_2CH_2Br$ as the minor product.

Mechanism

The driving force for the formation of $CH_3CHBrCH_3$ is the intermediate carbocation, shown in the mechanism below.

All students should know this mechanism but only AQA and WJEC demand an explanation in terms of carbocation stabilities.

The reaction with H_2SO_4 proceeds via a similar mechanism. H_2SO_4 can be treated as $H–OSO_2OH$.

(AQA and WJEC only)
The secondary carbocation is more stable with 2 alkyl groups feeding electrons towards the positive charge.
The alternative primary carbocation has only 1 alkyl group to stabilise the charge.

Addition polymerisation of alkenes

AQA	M2	OCR	M2
EDEXCEL	M1	SALTERS	M2
CCEA	M2	WJEC	M2

An addition polymer is a long-chain molecule with a high molecular mass, made by joining together many small molecules called monomers:

MANY MONOMERS $\longrightarrow$ SINGLE POLYMER

n molecules $\longrightarrow$ 1 single molecule

- The **monomer** used is an **unsaturated** alkene containing a double C=C bond.
- The addition **polymer** formed is a **saturated** compound **without** a double bond.

Many different addition polymers can be formed by using different alkene monomer units, based on the ethene molecule.

The equations below show the formation of poly(ethene), poly(chloroethene) (*pvc*) and poly(tetrafluoroethene) (*PTFE*).

You should be able to draw a short section of a polymer given the monomer units (and *vice versa*).

Make sure that you can show the repeat unit of a polymer.

- Notice that the principle is the same for each addition polymerisation.

- Each of the polymer structures shows the *repeat unit* in brackets. This is repeated thousands of times in each polymer molecule.

- It is important to show the repeat unit with 'side-links' to indicate that both sides are also attached to other repeat units.

tetrafluoroethene

poly(tetrafluoroethene) (PTFE)

Properties of monomers and polymers

- The monomers are volatile liquids or gases.
- Polymers are solids.
- This difference in physical properties is explained by increased van der Waals' forces between the much larger polymer molecules (see page 64).

Problems with disposal

- Addition polymers are non-biodegradable and take many years to break down.
- Disposal by burning can produce toxic fumes (e.g. depolymerisation produces monomers; dioxins from combustion of chlorinated polymers).

Rather than simply disposing of waste polymers, there is now movement towards recycling, combustion for energy production, and use as a feedstock for cracking.

This eliminates problems with disposal and helps to preserve finite energy resources.

Degradable polymers are also being developed from renewable resources such as corn starch.

Different types of polymer

SALTERS   M2

A **thermoplastic** is a polymer that becomes soft and plastic on heating and hard on cooling. This process **can** be repeated.

A **thermoset** is a polymer that sets permanently when heated. This process **cannot** be repeated.

A **co-polymer** is a polymer made by reacting together two different monomers.

Soluble polymers

SALTERS   M2

Hydrocarbon polymers such as poly(ethene) are non-polar and insoluble in water. Polymers with polar side-chains can have very different properties. For example, poly(ethenol), made by polymerisation of ethanol, is soluble in hot water.

Poly(ethenol) is used to manufacture plastic laundry bags for hospitals. The bags dissolve during the washing, protecting hospital workers from the risk of infection from handling dirty bedding.

ethenol

poly(ethenol)

Poly(ethenol) dissolves in water because the hydroxyl groups, –OH, are able to form hydrogen bonds to water molecules.

Progress check

1. Write the structural formula for the organic product for the reaction of but-2-ene with the following:
   (a) Br_2;   (b) H_2/Ni;   (c) HBr;   (d) H_2O/H_3PO_4.

2. Show the repeat unit for the addition polymer formed from propene.

1 (a) $CH_3CHBrCHBrCH_3$ (b) $CH_3CH_2CH_2CH_3$ (c) $CH_3CH_2CHBrCH_3$
(d) $CH_3CH_2CHOHCH_3$.

4.5 Alcohols

After studying this section you should be able to:

- *understand the polarity and physical properties of alcohols*
- *describe how ethanol is prepared*
- *describe oxidation reactions of alcohols*
- *describe nucleophilic substitution reactions of alcohols*
- *describe esterification and dehydration of alcohols*
- *describe tests for the presence of the hydroxyl group*

LEARNING SUMMARY

Alcohols General formula: $C_nH_{2n+1}OH$

AQA	M2	OCR	M2
EDEXCEL	M2	SALTERS	M2
CCEA	M2	WJEC	M2

Types and naming of alcohols

The functional group in alcohols is the **hydroxyl** group, C–OH.

Alcohols can be classified as primary, secondary or tertiary, depending on how many alkyl groups are bonded to C–OH. You can see how these types of alcohols are named in the diagrams below.

primary alcohol *1 alkyl group attached to C–OH*	*secondary alcohol* *2 alkyl groups attached to C–OH*	*tertiary alcohol* *3 alkyl groups attached to C–OH*
 propan-1-ol	 propan-2-ol	 methylpropan-2-ol

Polarity of alcohols

Carbon, oxygen and hydrogen have different electronegativities, and alcohols have polar molecules:

Oxygen is more electronegative than both carbon and hydrogen – both carbon and hydrogen are electron deficient.

The polarity produces electron-deficient carbon and hydrogen atoms, indicated in the diagram above. Depending upon the reagents used, alcohols can react by breaking either the C–O or O–H bond.

The properties of alcohols are dominated by the hydroxyl group, C–OH.

Physical properties of alcohols

The –OH group dominates the physical properties of short-chain alcohols.

Hydrogen bonding takes place between alcohol molecules, resulting in:

- higher melting and boiling points than alkanes of comparable M_r
- solubility in water.

hydrogen bonding between ethanol molecules

The solubility of alcohols in water decreases with increasing carbon chain length as the non-polar contribution to the molecule becomes more important.

miscible with water *large polar contribution*	*insoluble in water* *most of the molecule is a non-polar carbon chain*

Preparation of ethanol

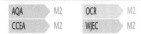

Ethanol finds widespread uses: in alcoholic drinks, as a solvent in the form of methylated spirits, and as a fuel. There are two main methods for its production.

Fermentation of sugars (for alcoholic drinks)

Yeast is added to an aqueous solution containing sugars.

E.g. $$C_6H_{12}O_6(aq) \longrightarrow 2C_2H_5OH(aq) + 2CO_2(g)$$

Hydration of ethene (for industrial alcohol)

Ethene and steam are passed over a phosphoric acid catalyst at 330°C under high pressure (see pages 131–132).

$$C_2H_4(g) + H_2O(g) \longrightarrow C_2H_5OH(l)$$

The preparation method used for ethanol may depend on the raw materials available.

An oil-rich country may use oil to produce ethene, then ethanol. However, this method uses up finite oil reserves.

In a country without oil reserves (especially in hot climates), sugar may provide a renewable source for ethanol production.

Combustion of alcohols

Alcohols such as ethanol and methanol are used as fuels, making use of combustion.

$$C_2H_5OH + 3O_2 \longrightarrow 2CO_2 + 3H_2O$$

Ethanol is used as a petrol substitute in countries with limited oil reserves.

Methanol is an 'oxygenate'. The oxygen in its formula aids combustion.

Methanol is used as a petrol additive in the UK to improve combustion of petrol. It also has increasing importance as a feedstock in the production of organic chemicals.

Oxidation of alcohols

The oxidation of different alcohols is an important reaction in organic chemistry. It links alcohols with aldehydes, ketones and carboxylic acids, shown below.

aldehyde, RCHO   ketone, RCOR'   carboxylic acid, RCOOH

- Aldehydes and ketones both contain the **carbonyl group** C=O.
- Carboxylic acids contain the **carboxyl** group COOH.

The stages of oxidation are shown below:

hydroxyl group C–OH $\longrightarrow$ carbonyl group C=O $\longrightarrow$ carboxyl group COOH

Oxidation of primary alcohols

A primary alcohol can be oxidised to an aldehyde and then to a carboxylic acid.

> **KEY POINT**
>
> Oxidation of an organic compound involves gain of oxygen OR loss of hydrogen (accompanied by loss of electrons from the organic compound).

The stages of oxidation are shown below:

primary alcohol   aldehyde   carboxylic acid

This is carried out using an oxidising agent. The oxidising agent used is a mixture of concentrated sulfuric acid, H_2SO_4 (source of H^+) and potassium dichromate, $K_2Cr_2O_7$ (source of $Cr_2O_7^{2-}$).

> For balanced equations, the oxidising agent can be shown simply as [O].

- By heating and distilling the product immediately, oxidation can be stopped at the aldehyde stage.

$$CH_3CH_2OH + [O] \longrightarrow CH_3CHO + H_2O$$

- By refluxing with an excess of the oxidising agent, further oxidation takes place to form the carboxylic acid.

$$CH_3CH_2OH + 2[O] \longrightarrow CH_3COOH + H_2O$$

Oxidation of secondary alcohols

> The orange dichromate ions, $Cr_2O_7^{2-}$, are reduced to green Cr^{3+} ions.

By heating with $H^+/Cr_2O_7^{2-}$ a secondary alcohol can be oxidised to form a ketone. No further oxidation normally takes place.

secondary alcohol $\longrightarrow$ ketone

secondary alcohol   ketone

The equation for the oxidation of propan-2-ol is shown below.

$$CH_3(CHOH)CH_3 + [O] \longrightarrow CH_3COCH_3 + H_2O$$

Oxidation of tertiary alcohols

Tertiary alcohols are not oxidised under normal conditions.

tertiary alcohol

Distinguishing between aldehydes and ketones (AQA only)

The presence of an aldehyde can easily be detected using two tests.

Heat aldehyde with Tollens' reagent

> Tollens' reagent is a solution of silver nitrate (as a source of Ag^+ ions) in ammonia.
> $Ag^+ \longrightarrow Ag$
> Fehling's (or Benedict's) solution is an alkaline solution of Cu^{2+} ions.
> $Cu^{2+} \longrightarrow Cu^+$

- Tollens' reagent $\longrightarrow$ silver mirror

Heat aldehyde with Fehling's (or Benedict's) solution

- Benedict's / Fehling's solution $\longrightarrow$ brick-red precipitate of Cu_2O

Ketones cannot be oxidised and **do not** react with the above solutions and this provides an easy method for distinguishing between an aldehyde and a ketone.

Identifying alcohols with the CH$_3$CHOH group (CCEA only)

The presence of an alcohol containing the CH_3CHOH group can be detected by heating an alcohol with iodine, I_2, in aqueous sodium hydroxide, $NaOH(aq)$.

> Alcohols containing the CH_3CHOH group produce pale-yellow crystals with $I_2/NaOH$.

- A pale-yellow crystalline solid called iodoform (triiodomethane), CHI_3, is formed. Iodoform has an antiseptic smell.

- Iodoform can also be confirmed following recrystallisation and determination of its melting point (123°C).

Substitution reactions of alcohols

EDEXCEL M2   SALTERS M2
CCEA M2

The electron-deficient carbon atom of the polar $C^{\delta+}–O^{\delta-}$ bond of alcohols attracts *nucleophiles*, allowing a **substitution** reaction to take place in which the nucleophile replaces the OH group. This is called **nucleophilic substitution** (see also halogenoalkanes, page 139).

> **KEY POINT**
>
> A nucleophile, such as OH^-, Br^-, H_2O , NH_3, CH_3OH, CN^-
> - is an electron-rich species
> - 'attacks' an electron-deficient carbon atom by **donating a pair of electrons**.

Reaction with bromide ions, Br⁻ in acid, H⁺

$$C_2H_5OH + Br^- + H^+ \longrightarrow C_2H_5Br + H_2O$$

To prepare a chloroalkane, use concentrated HCl.

For a bromoalkane or an iodoalkane, use:

Concentrated acid oxidises bromides and iodides to the halogens (see pages 82–83).

- a metal halide, e.g. NaBr
- and 50% sulfuric acid.

Reaction with phosphorus pentachloride PCl₅ (Edexcel and CCEA only)

Alcohols react with PCl_5, releasing hydrogen chloride gas, which forms misty fumes in air.

The reaction of alcohols with PCl_5 can be used as a test for an –OH group.

$$C_2H_5OH + PCl_5 \longrightarrow C_2H_5Cl + HCl + POCl_3$$

In the presence of ammonia, HCl fumes form dense white fumes of ammonium chloride:

(CCEA only)
A similar reaction takes place with thionyl chloride, $SOCl_2$, in which SO_2 forms instead of $POCl_3$.

$$NH_3 + HCl \longrightarrow NH_4Cl$$

Dehydration of alcohols

AQA M2   SALTERS M2
OCR M2   WJEC M2

When an alcohol is refluxed with a concentrated acid catalyst such as sulfuric acid H_2SO_4, or phosphoric acid H_3PO_4, an **elimination** reaction takes place: water is eliminated with formation of an alkene.

$$C_2H_5OH \longrightarrow C_2H_4 + H_2O$$

Use hot concentrated acid for elimination of water from an alcohol.

Elimination (1 ⟶ 2) is the opposite process to addition (2 ⟶ 1).

> **KEY POINT**
>
> An *elimination* reaction involves one species breaking up into two – the opposite to addition.
>
> Elimination usually converts a **saturated** compound into an **unsaturated** compound.

See page 135 for more details of methods used to prepare ethanol.

Alkenes, such as ethene, can be synthesised from renewable resources rather than from crude oil. Sugars, such as glucose, can be fermented to produce ethanol, C_2H_5OH.

E.g.

$$C_6H_{12}O_6 \longrightarrow 2C_2H_5OH + 2CO_2$$
glucose

Elimination of water from ethanol with an acid catalyst produces ethene.

$$C_2H_5OH \xrightarrow[\text{heat}]{\text{H}^+ \text{ catalyst}} C_2H_4 + H_2O$$

The alkenes can then be polymerised to form plastics.

Formation of esters

CCEA   M2
OCR   M2

Esters are formed by reaction of a carboxylic acid with an alcohol.

Esters are formed by reaction of an alcohol with a carboxylic acid in the presence of an acid catalyst (e.g. concentrated sulfuric acid).

$$\text{carboxylic acid} + \text{alcohol} \longrightarrow \text{ester} + \text{water}$$

The esterification of ethanol by ethanoic acid is shown below:

$$CH_3COOH + CH_3OH \longrightarrow CH_3COOCH_3 + H_2O$$

(CCEA only)

Esters can also be formed by reacting an alcohol with an acyl chloride at room temperature.

E.g.

$$CH_3COCl + C_2H_5OH \longrightarrow CH_3COOC_2H_5 + HCl$$
ethanoyl chloride

Reaction with sodium

EDEXCEL   M2
CCEA   M2

Sodium releases H_2 from any compound with a hydroxyl group: alcohols, carboxylic acids (and water).

Alcohols react with sodium, releasing hydrogen gas. The reaction involves breaking of the O–H bond of the alcohol.

$$2C_2H_5OH + 2Na \longrightarrow 2C_2H_5O^-Na^+ + H_2$$
sodium ethoxide

Progress check

1 Why does ethanol dissolve in water?

2 Write equations to show possible oxidations for the following:
   (a) butan-1-ol;   (b) butan-2-ol.

3 What are the 2 possible organic products of the reaction between butan-2-ol and concentrated sulfuric acid at 170°C?

3 $CH_3CH=CHCH_3$ and $CH_3CH_2CH=CH_2$
2 (a) $CH_3CH_2CH_2CH_2OH + [O] \longrightarrow CH_3CH_2CH_2CHO + H_2O$
  $CH_3CH_2CH_2CH_2OH + 2[O] \longrightarrow CH_3CH_2CH_2COOH + H_2O$
  (b) $CH_3CH_2CHOHCH_3 + [O] \longrightarrow CH_3CH_2COCH_3 + H_2O$
1 Alcohol –OH group forms H-bonds with water molecules.

4.6 Halogenoalkanes (Haloalkanes)

After studying this section you should be able to:

- understand the nature of the polarity in halogenoalkanes
- describe nucleophilic substitution reactions of halogenoalkanes
- explain the relative rates of hydrolysis of different halogenoalkanes
- describe elimination reactions of halogenoalkanes
- understand that reaction conditions can be used to control the products
- understand some common practical techniques used in organic chemistry

LEARNING SUMMARY

Halogenoalkanes General formula: $C_nH_{2n+1}X$, where X = F, Cl, Br or I

AQA	M2	OCR	M2
EDEXCEL	M2	SALTERS	M2
CCEA	M2	WJEC	M2

Uses of halogenoalkanes

Chlorofluorocarbons, CFCs were used as refrigerants, solvents, propellants, dry cleaning and degreasing agents, and for blowing polystyrene.

CFCs are now known to deplete the ozone layer. Their use is now banned in countries that have signed up to the Montreal Protocol.

Chloroethene and tetrafluoroethene are used to produce the plastics PVC and PTFE.

Halogenoalkanes are used as synthetic intermediates in chemistry and solvents.

Polychlorinated compounds are used as herbicides.

Types and naming of halogenoalkanes

1-bromopropane
primary

2-bromopropane
secondary

2-bromo-2-methylpropane
tertiary

Polarity of halogenoalkanes

Carbon and halogens have different electronegativities and halogenoalkanes have polar molecules with a polar C–X bond.

chlorine is more electronegative than carbon
electron flow from carbon to chlorine –
dipole produced

$$C^{\delta+}-F^{\delta-}$$
$$C^{\delta+}-Cl^{\delta-}$$ polarity
$$C^{\delta+}-Br^{\delta-}$$ **decreases**
$$C^{\delta+}-I^{\delta-}$$

The polarity produces an electron-deficient carbon atom, $C^{\delta+}$ which is important in the reactions of halogenoalkanes. The polarity decreases from fluorine to iodine, reflecting the decrease in electronegativity down the halogen group.

Nucleophilic substitution reactions of halogenoalkanes

AQA	M2	OCR	M2
EDEXCEL	M2	SALTERS	M2
CCEA	M2	WJEC	M2

The electron-deficient carbon atom of the polar $C^{\delta+}-X^{\delta-}$ bond attracts **nucleophiles**, allowing a **nucleophilic substitution** reaction to take place in which the nucleophile replaces the halogen atom (see also alcohols, page 137).

The hydrolysis of halogenoalkanes

> A **hydrolysis** reaction is a reaction in which water breaks a bond.
>
> KEY POINT

Reaction with aqueous hydroxide ions, $OH^-(aq)$ produces an alcohol.

$$C_2H_5Br + OH^-(aq) \xrightarrow[reflux]{OH^-/H_2O} C_2H_5OH + Br^-$$

The hydroxide ion, OH^-, behaves as a nucleophile.

Mechanism of nucleophilic substitution with primary halogenoalkanes

Primary halogenoalkanes react in a **one-step** mechanism.

Note the usefulness of the curly arrow in showing movement of an electron pair:

NaOH(aq) is used as a source of OH⁻ (aq).

The OH⁻ ion behaves as a nucleophile by donating an electron pair.

- the :OH⁻ nucleophile donates its electron pair to the electron-deficient $C^{\delta+}$ of the bromoalkane, forming a covalent bond;
- the C–Br bond is broken by heterolytic fission with loss of its electron pair to form a :Br⁻ ion.

Mechanism of nucleophilic substitution with tertiary halogenoalkanes
(CCEA only)

With tertiary halogenoalkanes, substitution with hydroxide ions takes place in a **two-step** mechanism.

carbocation intermediate

- In the 1st step, a Br⁻ ion is lost.
- The carbocation intermediate is stabilised by the three alkyl groups (see also page 132).
- In the 2nd step, nucleophilic attack by a hydroxide ion forms a tertiary alcohol.

Comparing the hydrolysis reactions of halogenoalkanes

The rates of hydrolysis of different halogenoalkanes can be investigated as follows:

- Alkaline hydrolysis with NaOH(aq)/H_2O, reflux:

$$R–X + OH^- \longrightarrow R–OH + X^-$$

An alternative method uses water itself for the hydrolysis. The halogenoalkane is heated with a mixture of water, ethanol and $AgNO_3$(aq). The rate of hydrolysis is monitored directly by observing precipitation of replaced halide ions as silver halides.

- Acidification with dilute nitric acid. This removes excess NaOH, which would otherwise form a precipitate with the silver nitrate in the next stage, preventing detection of any silver halides.
- Addition of $AgNO_3$(aq) shows the presence of aqueous halide ions (see also Testing for halide ions, page 82).

$$Ag^+(aq) + X^-(aq) \longrightarrow AgX(s)$$

The intensity of any precipitate indicates the extent and rate of the hydrolysis.

$C^{\delta+}–F^{\delta-}$
$C^{\delta+}–Cl^{\delta-}$ polarity
$C^{\delta+}–Br^{\delta-}$ **increases**
$C^{\delta+}–I^{\delta-}$

bond	bond enthalpy /kJmol⁻¹
C–F	467
C–Cl	364
C–Br	290
C–I	228

Two factors – which is more important?

- **Polarity** predicts that the more polar C–F bond would attract water most easily and would give the fastest reaction.
- **Bond enthalpy** predicts that the C–I bond would be broken most easily and would give the fastest reaction.

For this reaction, bond enthalpy is more important than polarity.
The rate of hydrolysis increases as the C–X bond weakens.

KEY POINT

Further examples of nucleophilic substitution

| AQA | M2 | CCEA | M2 |
| EDEXCEL | M2 | SALTERS | M2 |

Nucleophilic substitution of halogenoalkanes is used in organic synthesis to introduce different functional groups onto the carbon skeleton. The reaction scheme below shows how halogenoalkanes react with ammonia and cyanide ions.

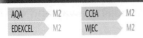

For hydrolysis, water is used as the solvent. For other nucleophilic substitutions, it is important that water is absent or hydrolysis will take place.

Ethanol is used as the alternative solvent.

$$H-\overset{\underset{\displaystyle H}{|}}{\underset{\underset{\displaystyle H}{|}}{C}}-\overset{\underset{\displaystyle H}{|}}{\underset{\underset{\displaystyle H}{|}}{C}}-Br$$

(AQA and CCEA only)

NH₃/ethanol, reflux

CN⁻/ethanol, reflux

$$H-\overset{\underset{\displaystyle H}{|}}{\underset{\underset{\displaystyle H}{|}}{C}}-\overset{\underset{\displaystyle H}{|}}{\underset{\underset{\displaystyle H}{|}}{C}}-NH_2 \quad + \quad HBr$$

AMINE

(Note that HBr then reacts:
$$NH_3 + HBr \longrightarrow NH_4Br)$$

$$H-\overset{\underset{\displaystyle H}{|}}{\underset{\underset{\displaystyle H}{|}}{C}}-\overset{\underset{\displaystyle H}{|}}{\underset{\underset{\displaystyle H}{|}}{C}}-CN \quad + \quad Br^-$$

NITRILE

The ability of halogenoalkanes to undergo nucleophilic substitution is important in organic synthesis as compounds with different functional groups can be synthesised.

Elimination reactions of halogenoalkanes

| AQA | M2 | CCEA | M2 |
| EDEXCEL | M2 | WJEC | M2 |

When a halogenoalkane is refluxed with hydroxide ions in **anhydrous** conditions (using NaOH in **ethanol** as a solvent at 78°C), an **elimination** reaction takes place.

$$H-\overset{\underset{\displaystyle H}{|}}{\underset{\underset{\displaystyle H}{|}}{C}}-\overset{\underset{\displaystyle H}{|}}{\underset{\underset{\displaystyle H}{|}}{C}}-Br \quad + \quad OH^- \quad \xrightarrow[\text{reflux}]{OH^-/\text{ethanol}} \quad \overset{H}{\underset{H}{>}}C=C\overset{H}{\underset{H}{<}} \quad + \quad Br^- \quad + \quad H_2O$$

The hydroxide OH⁻ ion behaves as a base, accepting a proton H⁺ to form H_2O.

Elimination versus hydrolysis

| AQA | M2 | CCEA | M2 |
| EDEXCEL | M2 | WJEC | M2 |

The reaction of hydroxide ions with halogenoalkanes forms different organic products, depending upon the reaction conditions used.

Use different reaction conditions to control the type of reaction.

Commonly tested in exams.

> **KEY POINT**
>
> aqueous conditions ⟶ alcohol: OH⁻ acts as a **nucleophile**
> anhydrous conditions ⟶ alkene: OH⁻ acts as a **base**

The ability to control *which* reaction takes place solely by changing the conditions is an important tool for the synthetic chemist.

Progress check

1 What is meant by the term nucleophilic substitution?

2 Why does 1-bromopropane react with nucleophiles but propane does not?

3 Why do bromoalkanes react more readily than chloroalkanes?

4 What are the organic products for the reactions of 1-bromopropane with:
(a) NaOH/H₂O;   (b) NH₃/ethanol;   (c) NaOH/ethanol?

4 (a) CH₃CH₂CH₂OH (b) CH₃CH₂CH₂NH₂ (c) CH₃CH=CH₂.
3 C–Br bond is weaker than C–Cl bond.
2 C–Br bond is polar, C–H bonds are not polar.
1 Replacement of an atom or group in an organic molecule by a nucleophile.

141

Organic practical techniques

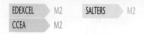

| EDEXCEL | M2 | SALTERS | M2 |
| CCEA | M2 | | |

Common techniques

During your AS Chemistry course, you may have carried out one or more of the following experiments.

- The preparation of a chloroalkane from an alcohol using HCl.
- The oxidation of alcohols to form an aldehyde or a carboxylic acid.
- Preparation of cyclohexene by dehydration of cyclohexanol.

Stages in the preparation of a pure organic liquid product

You will have developed some important practical skills. You may be assessed directly on your practical skills but you may also be asked about some of the following techniques on theory papers.

Important organic practical skills

- **Heating** under **reflux**.
  – The boiling of a liquid and subsequent condensation back to the same container. This prevents the organic solvent from boiling away and the heated reactants from drying out.
- **Separating** immiscible liquids using a separating funnel.
  – Each liquid can be run off, in turn, through the tap.
- **Removing impurities** using a separating funnel.
  – **Acidic impurities** can be removed by shaking an organic liquid with sodium hydrogencarbonate solution, which reacts with any acid.
- **Drying** an organic liquid.
  – An anhydrous salt is used such as anhydrous sodium sulfate.
- **Distillation** of an organic liquid.
  – Distillation allows collection of a pure liquid product. You can separate organic liquids in a mixture by distilling each off at its boiling point (fractional distillation).

> Remember to take the stopper out or nothing will run out of the tap!

reflux

separating immiscible liquids

distillation

4.7 Analysis

After studying this section you should be able to:

- understand that infra-red spectroscopy can be used to identify functional groups
- interpret a simple infra-red spectrum
- use a mass spectrum to determine the relative molecular mass of an organic molecule
- use the fragmentation pattern of a mass spectrum to deduce likely structures for an unknown compound

LEARNING SUMMARY

Using infra-red (IR) spectroscopy in analysis

AQA	M2	OCR	M2
EDEXCEL	M2	SALTERS	M2
CCEA	M2	WJEC	M2

AQA (A2)	M4	OCR (A2)	M4
EDEXCEL (A2)	M4, M5	SALTERS (A2)	M4, M5
		WJEC (A2)	M4

Basic principles

The frequency of IR absorption is measured in wavenumbers, units: cm⁻¹.

Bonds in molecules naturally vibrate. Some bonds in molecules increase their vibrations by absorbing energy from IR radiation. Different bonds absorb different frequencies of IR radiation.

An IR spectrum is obtained by passing a range of IR frequencies through a compound. As energy is taken in, **absorption peaks** are produced. The frequencies of the absorption peaks can be matched to those of known bonds to identify structural features in an unknown compound.

IR radiation has less energy than visible light.

> IR spectroscopy is useful for identifying the functional groups in a molecule.
>
> KEY POINT

Important IR absorptions

You don't need to learn the absorption frequencies – the data is provided.

bond	functional group	wavenumber/cm⁻¹
O–H	hydrogen bonded in alcohols	3230 – 3550
N–H	amines	3100 – 3500
C–H	organic compound with a C–H bond	2850 – 3100
O–H	hydrogen bonded in carboxylic acids	2500 – 3300 (broad)
C≡N	nitriles	2200 – 2260
C=O	aldehydes, ketones, carboxylic acids, esters	1680 – 1750
C–O	alcohols, esters	1000 – 1300

IR spectroscopy is used in some modern breathalysers for measuring the concentration of blood alcohol. A particular IR absorption identifies the presence of ethanol in the breath and the intensity of the peak is directly related to the ethanol level.

An infra-red spectrum is particularly useful for identifying:

- an **alcohol** from absorption of the O–H bond
- a **carbonyl** compound from absorption of the C=O bond
- a **carboxylic acid** from absorption of the C=O bond **and** broad absorption of the O–H bond.

Interpreting infra-red spectra

AQA	M2	OCR	M2
EDEXCEL	M2	SALTERS	M2
CCEA	M2	WJEC	M2

AQA (A2)	M4	OCR (A2)	M4
EDEXCEL (A2)	M4, M5	SALTERS (A2)	M4, M5
		WJEC (A2)	M4

Carbonyl compounds (aldehydes and ketones)
Butanone,
$CH_3COCH_2CH_3$

- C=O absorption 1680 to 1750 cm⁻¹

C=O absorption at 1740 cm⁻¹

Alcohols

Ethanol,

C_2H_5OH

- O–H absorption 3230 to 3500 cm^{-1}
- C–O absorption 1000 to 1300 cm^{-1}

> IR spectroscopy is most useful for identifying C=O and O–H bonds.
> Look for the distinctive patterns.

O–H absorption at 3300 cm^{-1}

C–O absorption at 1010 cm^{-1}

wavenumber / cm^{-1}

Carboxylic acids

Propanoic acid,

C_2H_5COOH

- Very broad O–H absorption 2500 to 3500 cm^{-1}
- C–O absorption 1680 to 1750 cm^{-1}

> Note that all these molecules contain C–H bonds, which absorb in the range 2840 to 3045 cm^{-1}.

Very broad O–H absorption between 2500 and 3500 cm^{-1}

C=O absorption at 1680 cm^{-1}

C–O absorptions at 1200 and 1030 cm^{-1}

wavenumber / cm^{-1}

Ester

Ethyl ethanoate,

$CH_3COOC_2H_5$

- C=O absorption 1680 to 1750 cm^{-1}
- C–O absorption 1000 to 1300 cm^{-1}

> There are other organic groups (e.g. N–H, C=C) that absorb IR radiation but the principle of linking the group to the absorption wavenumber is the same.

C=O absorption at 1740 cm^{-1}

C–O absorptions at 1220 and 1020 cm^{-1}

wavenumber / cm^{-1}

> The fingerprint region is unique for a particular compound.

Fingerprint region

- Between 1000 and 1550 cm^{-1}

Many spectra show a complex pattern of absorption in this range.

- This pattern can allow the compound to be identified by comparing its spectrum with spectra of known compounds.

Progress check

1 Ethanol, CH_3CH_2OH was oxidised to ethanal, CH_3CHO and then to ethanoic acid, CH_3COOH. How could you use IR spectroscopy to follow the course of this reaction?

1 Ethanol absorbs at about 3300 cm^{-1} (OH); ethanal absorbs at about 1700 cm^{-1} (C=O); ethanoic acid absorbs at about 1700 cm^{-1} (C=O) and 2500–3500 cm^{-1} (broad) (OH from COOH group).

Mass spectrometry

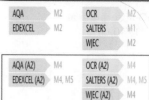

AQA	M2	OCR	M2
EDEXCEL	M2	SALTERS	M1
		WJEC	M2

AQA (A2)	M4	OCR (A2)	M4
EDEXCEL (A2)	M4, M5	SALTERS (A2)	M4, M5
		WJEC (A2)	M4

Mass spectrometry can be used to determine relative atomic masses from a mass spectrum (see page 31). Mass spectrometry is also used to determine relative molecular masses and to identify the molecular structures of organic compounds.

Molecular ions

The molecular ion peak is usually given the symbol M.

Organic molecules can be analysed using mass spectrometry.

In the mass spectrometer, organic molecules are bombarded with electrons. This can lead to the formation a **molecular ion**.

The mass spectrum also contains fragments of the molecule ion (see detail below).

The equation below shows the formation of a molecular ion from butanone, $CH_3COCH_2CH_3$.

$$H_3C-\overset{\overset{\displaystyle O}{\|}}{C}-CH_2CH_3 \ + \ e^- \ \longrightarrow \ \left[H_3C-\overset{\overset{\displaystyle O}{\|}}{C}-CH_2CH_3\right]^+ \ + \ 2e^-$$

molecular ion, *m/z*: 72

- The molecular ion peak, M, provides the relative molecular mass of the compound.

High resolution mass spectrometry

AQA   M2

SALTERS (A2)   M4   CCEA (A2)   M5

Molecules of similar relative molecular masses can be distinguished using high resolution mass spectrometry.

Using this technique:

- the molecular mass is obtained with a very accurate value
- this is compared with very accurate relative isotopic masses.

Worked Example

isotope	relative isotopic mass
^1H	1.0078
^{12}C	12.0000
^{14}N	14.0031
^{16}O	15.9949

Three gases, CO, C_2H_4 and N_2, each have an approximate relative molecular mass of 28.

The table below shows accurate molecular masses for these gases, calculated from accurate relative isotopic masses (see the margin).

gas	CO	C_2H_4	N_2
molecular mass	27.9949	28.0312	28.0062

The gases can be distinguished by matching the *m/z* values from high resolution mass spectrometry with the calculated values.

Fragment ions

EDEXCEL   M2   OCR   M2

WJEC   M2

AQA (A2)   M4   OCR (A2)   M4
EDEXCEL (A2)   M4, M5   SALTERS (A2)   M4, M5
   WJEC (A2)   M4

In the conditions within the mass spectrometer, some molecular ions are fragmented by bond fission.

The fragmentation pattern provides clues about the molecular structure of the compound.

- The bond fission that takes place is a fairly random process:
  - different bonds are broken
  - a **mixture of fragment ions** is obtained.
- The mass spectrum contains both the molecular ion and the mixture of fragment ions.

Mass spectrometry of organic compounds is useful for identifying:

- the relative molecular mass
- parts of the skeletal formula.

Fragmentation of butane, C_4H_{10}

The mass spectrum of butane would contain peaks for these four ions:

$C_4H_{10}{}^+$: *m/z* = 58
$C_3H_7{}^+$: *m/z* = 43
$C_2H_5{}^+$: *m/z* = 29
$CH_3{}^+$: *m/z* = 15

Bond fission forms a fragment ion and a radical. The diagram below shows that fission of the same C–C bond in a butane molecule can form two different fragment ions.

$H_3C-CH_2-CH_2\!-\!\!CH_3{}^+$
m/z = 58

loss of 15 mass units → $H_3C-CH_2-CH_2{}^+$ + $^\bullet CH_3$
m/z = 43

loss of 43 mass units → $H_3C-CH_2-\overset{\bullet}{C}H_2$ + $CH_3{}^+$
m/z = 15

- Fragmentation of the molecular ion produces a fragment ion by loss of a radical.
- The mass spectrometer only detects the fragment ion.
- The fragment ion may itself fragment into another ion and radical.

The mass spectrum of butanone

> Note that only ions are detected in the mass spectrum.
>
> Uncharged species such as the radical cannot be deflected within the mass spectrometer.

The mass spectrum of butanone, $CH_3CH_2COCH_3$, is shown below. The m/z values of the main peaks have been labelled.

> The M+1 peak is a small peak 1 unit higher than the molecular ion peak.
>
> The origin of the M+1 peak is the small proportion of carbon-13 in the carbon atoms of organic molecules.

The table below shows the identities of the main peaks.

> A common mistake in exams is to show both the fragmentation products as ions.

m/z	ion	fragment lost
72	$CH_3COCH_2CH_3^+$	–
57	$CH_3CH_2CO^+$	$CH_3\bullet$
43	CH_3CO^+	$CH_3CH_2\bullet$
29	$CH_3CH_2^+$	$CH_3CO\bullet$
15	CH_3^+	$CH_3CH_2CO\bullet$

> The acylium (RCO^+) ion is particularly stable and is often present as a high dominant peak.

The most abundant peak in the mass spectrum is the base peak.

- The base peak in the mass spectrum of butanone has m/z: 43, formed by the loss of a $CH_3CH_2\bullet$ radical:

> The base peak is given a relative abundance of 100. Other peaks are compared with this base peak.

$$\left[H_3C-\overset{\overset{O}{\|}}{C}-CH_2CH_3 \right]^+ \longrightarrow \left[H_3C-\overset{\overset{O}{\|}}{C} \right]^+ + \bullet CH_2CH_3$$

molecular ion	*fragment ion*	*radical*
m/z: 72	m/z: 43	M – 29

Common patterns in mass spectra

Different fragmentations are possible depending on the structure of the molecular ion.

- The table above shows common fragment ions and fragmented radicals.
- The skill in interpreting a mass spectrum is in searching for known patterns, evaluating all the evidence and reassembling all the information into a molecular structure.
- You should also look for a peak at m/z 77 or loss of 77 units. This is a giveaway for the presence of a phenyl group, C_6H_5, in a molecule.

Progress check

1 Write an equation for the formation of the fragment ion at $m/z=29$ in the mass spectrum of butanone above.

2 What peaks would you expect in the mass spectrum of pentane?

2 m/z: 72, $C_5H_{12}^+$; m/z: 57, $C_4H_9^+$; m/z: 43, $C_3H_7^+$; m/z: 29, $C_2H_5^+$; m/z: 15, CH_3^+.

1 $CH_3CH_2COCH_3^+ \longrightarrow CH_3CH_2^+ + CH_3CO\bullet$

4.8 Chemistry in the environment

After studying this section you should be able to:

- *outline the causes of global warming and strategies to curtail the production of greenhouse gases*
- *explain the formation of the ozone layer and its depletion by radicals*
- *outline the main principles of green chemistry*

LEARNING SUMMARY

Global warming

| AQA | M2 | OCR | M2 |
| EDEXCEL | M2 | SALTERS | M2 |

The main gases in the air are:

nitrogen 78%

oxygen 21%

argon 0.9%

carbon dioxide 0.04%

Although O_2 and N_2 are the main gases in the atmosphere, their molecules do not change polarity when they vibrate and so they do not absorb IR radiation.

The greenhouse effect

The **greenhouse effect** is a natural process, keeping the Earth at a temperature capable of supporting life. Most IR radiation emitted by the Earth goes back into space. However, greenhouse gases in the atmosphere absorb some of the IR radiation, which is then re-emitted as heat, with some passing back towards the Earth.

- The Earth absorbs energy at the same rate as it radiates energy.
- The equilibrium maintains a steady temperature.

The absorption-emission process keeps the heat close to the surface of the Earth.

Greenhouse gases

Absorption of IR radiation by **greenhouse gases** causes bonds to vibrate. A molecule only absorbs IR radiation if its polarity changes as the bonds vibrate.

The main absorbers of IR radiation are the C=O, O–H and C–H bonds in H_2O, CO_2 and CH_4. Other IR absorbers include the nitrogen–oxygen bonds in nitrogen oxides such as NO and NO_2 (often referred to as NO_x).

The greenhouse effect of a gas depends on:

- its atmospheric concentration
- its ability to absorb IR radiation.

Climate change

Climate change refers to variations in the Earth's climate over time.
- Over hundreds of thousands of years, natural climate changes have been responsible for ice-ages, followed by warming.
- Over shorter periods of time, natural climate change can be the result of factors such as sunspot activity and the presence of volcanic dust in the atmosphere.

The Earth has always experienced **natural climate** change.

Scientific evidence has shown that substantial climate change is now taking place and that the Earth's climate is warming. Scientists agree that the production of greenhouse gases from **human activities** is the primary factor driving **climate change**.

- Human activity is producing **more** greenhouse gases, particularly CO_2.
- Increased proportions of greenhouse gases are upsetting the sensitive, natural balance of nature, resulting in man-made global warming.

Climate change derived from human activities is known as **anthropogenic** climate change, as opposed to natural climate change without human influences.

Computer modelling has predicted significant problems, e.g. melting ice caps, rising sea levels and more extreme storms and flooding.

International treaties, such as the Kyoto Protocol, have set targets for reductions in CO_2 production.

Carbon neutrality

AQA ▶ M2   OCR ▶ M2

Carbon neutrality can be achieved in various ways, for example:

changing from finite fuels, such as oil, to renewable fuels based on plants that can be grown annually, such as ethanol and rape-seed oil

cutting down fewer trees and growing more trees to take in CO_2

using less fuel and more public transport

carbon storage and capture (CSS)

improving the efficiency of engines.

The proportion of CO_2 in the atmosphere has increased more in the last hundred years than at any other time in the Earth's history. This increase is blamed on human activities. Carbon neutrality is seen as an important step in curbing global warming by re-balancing the CO_2 in the atmosphere.

> **KEY POINT**
>
> **Carbon neutral** refers to 'an activity that has no net annual carbon (greenhouse gas) emissions to the atmosphere'.

For carbon neutrality, amounts of CO_2 released into the atmosphere by human activities, such as from burning fuels, is balanced by the **same amounts** of CO_2 being taken in **from** the atmosphere during plant growth.

Alternatively, CO_2 produced during combustion could be stored or 'sequestered' by **carbon storage and capture** (CSS). This can achieved by:
- removal of waste carbon dioxide as a liquid, injected deep in the oceans
- storage as a gas in deep geological formations
- reacting carbon dioxide with metal oxides to form stable carbonate minerals.

This *trapping* of CO_2 emissions would substantially reduce CO_2 emissions from power stations. CSS could make the combustion of fossil fuels carbon neutral.

For transport, carbon neutrality can be approached by use of **biofuels**.

Although biofuels are often assumed to be carbon neutral, this is not strictly true – additional energy is required for transportation of the fuels, production of fertilisers, etc.

> **KEY POINT**
>
> A **biofuel** is a liquid or gas transportation fuel derived from biomass.

- Biofuels are **renewable**. Once used, fresh fuel can be made again from plants, grown in a short length of time.
- **Finite** resources, such as oil and coal take millions of years to form. Once used, fossil fuels are spent – they are non-renewable.

The best sources of biofuels are alcohols, such as ethanol and methanol, and vegetable oils, such as the rape-seed oil. Biofuels can be used with petrol to fuel cars.

The carbon neutrality of bio-ethanol is achieved in the following way:
- Growing plants produces sugars, taking in CO_2 from the atmosphere.
- The sugars are fermented with yeast, making bio-ethanol.
- The bio-ethanol is burned as a fuel, releasing CO_2 into the atmosphere.
- More sugars are made to form fresh bio-ethanol.

Carbon footprint

As with carbon neutrality, 'carbon footprint' is used to quantify CO_2 emissions.

Carbon footprint is the total amount of CO_2 and other greenhouse gases emitted over the full life cycle of a product or service.

Alternative fuels, such as hydrogen, would produce no CO_2 emissions but the processes required to produce H_2 as a fuel are currently energy intensive. Hydrogen fuels remain a goal for the future, particularly as a fuel for cars.

The ozone layer

EDEXCEL ▶ M2   OCR ▶ M2
SALTERS ▶ M2

The ozone layer is present 10–60 km above the Earth's surface in the region of the upper atmosphere known as the **stratosphere**.

The ozone layer forms naturally by the interaction of **ultraviolet** (UV) radiation with **oxygen**.

Ultraviolet radiation breaks O_2 molecules into oxygen radicals.

$$O_2 \longrightarrow 2O$$

The O radicals react with O_2 molecules to form **ozone** molecules, O_3. The bond-forming process releases energy as heat.

$$O_2 + O \longrightarrow O_3 + \text{heat}$$

O_3 molecules in turn absorb more UV radiation and are themselves broken into O_2 and O. An equilibrium is set up:

$$O_2 + O \rightleftharpoons O_3$$

The Sun mainly radiates radiation in the visible region of the spectrum, with an appreciable amount of UV.

The Earth mainly radiates IR radiation.

- The process maintains a natural concentration of ozone.
- Overall, **UV radiation is absorbed** and **heat is evolved**. This increases the temperature in the stratosphere.

The high energy of the UV radiation would cause harm to life on Earth. Excess UV can lead to skin cancer. The ozone layer is essential in protecting life on Earth by filtering out the harmful UV radiation.

Depletion of ozone

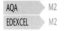

The ozone layer is beneficial for life on Earth.

Depletion of ozone is caused by **radicals**, particularly Cl and NO.

Cl and NO radicals have been formed from human activity, affecting the natural balance that maintains the ozone layer.

Cl radicals are formed by the action of **UV radiation** on **CFCs**, which were once used in aerosols and as refrigerants.

e.g. $CClF_3 \longrightarrow Cl + CF_3$

Oxides of nitrogen (NO_x) are formed from thunderstorms and aircraft flying through the upper atmosphere.

e.g. $N_2 + O_2 \longrightarrow 2NO$

Both processes lead to depletion of ozone in the stratosphere:

The Cl and NO radicals are recycled in chain reactions, which continue many thousand of times until two radicals collide, terminating the reaction. A single Cl radical can remove thousands of O_3 molecules from the ozone layer.

See also homogeneous catalysis: page 109.

stage 1 $O_3 + Cl\bullet \longrightarrow ClO\bullet + O_2$
stage 2 $ClO\bullet + O \longrightarrow O_2 + Cl\bullet$
overall: $O_3 + O \longrightarrow 2O_2$

stage 1 $O_3 + NO\bullet \longrightarrow NO_2\bullet + O_2$
stage 2 $NO_2\bullet + O \longrightarrow O_2 + NO\bullet$
overall: $O_3 + O \longrightarrow 2O_2$

In the processes above, Cl and NO act as catalysts.
- They are used up in the first stage.
- They are regenerated in the second stage.

Overall, Cl and NO radicals are **not consumed**.

Benefits of CFC use versus damage to ozone

In the 1930s, CFCs were seen as being the ideal chemicals for refrigerants, aerosols and as blowing agents for 'expanded poly(styrene)'.

CFCs have the ideal properties for these uses: high volatility, non-toxicity, non-flammability, no smell and extreme unreactivity.

This unreactivity causes the problem.

- Most pollutants are broken down naturally in the lower atmosphere. However, unreactive CFCs diffuse over many years up to the upper atmosphere.
- High-energy UV radiation in the stratosphere finally breaks down the CFCs, forming Cl radicals. The highly reactive Cl radicals attack the ozone.

In the 1970s and 1980s, scientists discovered that Cl radicals were responsible for thinning of the ozone layer.

- Most governments have signed up to the Montreal Protocol, which sets targets for reduced use of CFCs.
- Since the early 2000s, the use of CFCs has been banned internationally. Although replacement chemicals are now in use, many old refrigerators contain CFCs and these have to be disposed of carefully by collecting the CFCs.

It may take a century before the ozone layer fully recovers.

Green chemistry

EDEXCEL ▷ M2   SALTERS ▷ M2
OCR ▷ M2   WJEC ▷ M1

> Exam questions are likely to be set in the context of green chemistry. You need to know the main principles and be able to apply them.

The chemical industry is currently reinventing itself to make it more sustainable ('greener'). Green chemistry encourages the design of products and processes that reduce or eliminate the use and generation of hazardous substances. It seeks to produce more efficient processes with less energy input.

The main principles of green chemistry are listed below.

- Using industrial processes that reduce or eliminate hazardous chemicals and which involve the use of fewer chemicals.
  - Lead has been eliminated from petrol, paints and electrical components.
  - Waste CO_2 has replaced CFCs as the blowing agent in expanded polystyrene.
- Designing processes with a high atom economy (see page 124):
  - reducing waste and preventing pollution of the environment
  - discovering catalysts for improving the atom economies of reactions.
- Ensuring that any waste products produced are non-toxic:
  - producing waste that can be safely biodegraded into harmless substances
  - increasing the use of recycling.
- Consuming less energy:
  - replacing or supplementing fossil fuels with renewable fuels
  - making efficient use of energy or developing renewable energy sources.

Other factors

The apparent benefits may be offset by unexpected and detrimental side-effects.

Although rich countries may benefit from the use of renewable fuels, the population in the poorer countries may get insufficient food, e.g. the production of biodiesel uses grain crops and land.

Progress check

1 State three greenhouse gases and the bonds responsible for IR absorption.

2 Outline how the use of bio-ethanol approaches carbon neutrality.

3 Outline the natural processes responsible for maintaining the ozone layer in the stratosphere.

4 Write equations to show how the presence of CFC in the stratosphere depletes the ozone layer.

1 H_2O (O–H); CO_2 (C=O); CH_4 (C–H).

2 Growing plants produces sugars, taking in CO_2 from the atmosphere.
The sugars are fermented with yeast, making bio-ethanol.
The bio-ethanol is burned as a fuel, releasing CO_2 into the atmosphere
Overall, the CO_2 taken in during growth = CO_2 released to the atmosphere (although there will be some extra CO_2 released from transportation and fertiliser production.)

3 Ultraviolet radiation breaks O_2 molecules into oxygen radicals.
$O_2 \longrightarrow 2O$
The O radicals react with O_2 to form O_3, with excess energy released as heat.
$O_2 + O \longrightarrow O_3 + heat$
O_3 molecules absorb more UV radiation and are broken into O_2 and O.
An equilibrium is set up: $O_2 + O \rightleftharpoons O_3$

4 $CClF_3 \longrightarrow Cl + CF_3$
stage 1 $O_3 + Cl \bullet \longrightarrow ClO \bullet + O_2$
stage 2 $ClO \bullet + O \longrightarrow O_2 + Cl \bullet$
overall: $O_3 + O \longrightarrow 2O_2$

Sample question and model answer

For open-ended questions of this type, there are usually more marking points than the total – you have some breathing space.

1

(a) Explain, using examples, the following terms used in organic chemistry.

(i) *general formula*

This is the simplest algebraic representation for any member of a series of organic compounds, e.g. alkanes: C_nH_{2n+2} ✓

(ii) *homologous series*

This is a series containing compounds with similar chemical properties ✓ with each successive member differing by CH_2 ✓, e.g. 'alkanes' ✓

(iii) *homolytic fission* and *heterolytic fission*

Do make sure that you choose valid examples. In the actual examination, many candidates managed to confuse homolytic and heterolytic fission – poor preparation!

Fission means the breaking of a bond. ✓

In homolytic fission, one electron from the bond is released to each atom with radicals forming. ✓ e.g. $Cl_2 \longrightarrow 2Cl\bullet$ ✓

In heterolytic fission, both electrons in the bonding pair are released to one of the atoms with ions forming ✓ e.g. $Br_2 \longrightarrow Br^+ + Br^-$ ✓

9 marks $\longrightarrow$ [7]

Always explain your working. This type of problem is often easier than it looks.

You should look back at the question many times to check that you have answered all parts.

(b) Cracking of the unbranched compound **A**, C_6H_{14}, produced the saturated compound **B** and the unsaturated hydrocarbon **C** (M_r, 42). Compound **B** reacted with bromine in UV light to form a monobrominated compound **D** and an acidic gas **E**. Compound **C** reacted with hydrogen bromide to form a mixture of two compounds **F** and **G**.

(i) Use this evidence to suggest the identity of each of compounds **A** to **G**. Include equations for the reactions in your answer.

A is unbranched. ∴ **A** = $CH_3CH_2CH_2CH_2CH_2CH_3$ ✓

C has M_r = 42; must have a double bond: must contain 3 carbons
∴ **C** = C_3H_6. ✓

Notice that everything 'works': questions are designed in this way. If you find yourself with 'strange' structures containing functional groups that you haven't seen before, you have probably made a mistake. This is the time to check back through your logic.

Cracking of **A** produces an alkane and an alkene, **B** is an alkane: **B** = C_3H_8 ✓

The equation is: $C_6H_{14} \longrightarrow C_3H_6 + C_3H_8$ ✓

Alkanes react with bromine in UV by substitution. ∴ **D** = C_3H_7Br ✓,
E = HBr ✓

The equation is: $C_3H_8 + Br_2 \longrightarrow C_3H_7Br + HBr$ ✓

Alkenes react with bromine by addition. HBr can be added across a double bond in two ways to give: **F** and **G** as $CH_3CH_2CH_2Br$ ✓ and $CH_3CHBrCH_3$. ✓

The equation is: $C_3H_6 + HBr \longrightarrow C_3H_7Br$ ✓

(ii) Oil companies often process compounds such as **A** into better fuels. Suggest two organic products of this processing of **A**.

Any 2 examples here formed from C_6H_{14}. These could be branched hydrocarbons, aromatic or alicyclic – but any chosen **must** contain 6 carbons. (Read the question!)

Compound **A**, hexane, could be converted into benzene ✓ and 2-methylpentane. ✓

(iii) Predict the structure of the polymer that could be formed from compound **C**.

When drawing a polymer: 'side bonds' are always needed. Always show the repeat unit.

$$\left[\begin{array}{c} \underset{|}{\overset{H}{\underset{|}{C}}} - \overset{CH_3}{\underset{|}{\underset{|}{C}}} \\ \overset{|}{H} \quad \overset{|}{H} \end{array} \right]_n$$

✓✓(1 for no double bond, 1 for correct skeleton) [14]

Organic chemistry, analysis and the environment

Practice examination questions

1 There are four structural isomers with the molecular formula $C_3H_6Cl_2$.

   (a) (i) Explain the term *structural isomers*.

   (ii) Show the structural formulae of the structural isomers of $C_3H_6Cl_2$.

   [5]

   (b) Some propane was reacted with chlorine to form a mixture of isomers.

   (i) What conditions are required for this reaction to take place?

   (ii) Two structural isomers, **A** and **B**, were separated from the mixture. These isomers had a molar mass of 78.5 g mol^{-1}. Deduce the molecular formula of these two isomers.

   (iii) Draw the structural formulae of **A** and **B** and name each compound.

   [6]
   [Total: 11]

2 1,2-Dichloroethene has *E* and *Z* isomers.

   (a) (i) Explain the two conditions required for an organic molecule to have *E* and *Z* isomers.

   (ii) What is the difference between *E* and *Z* isomers?

   (iii) Draw and label the *E* and *Z* isomers of 1,2-dichloroethene.

   [5]

   (b) The *E* and *Z* isomers of 1,2-dichloroethene can both be reduced to 1,2-dichloroethane. What are the reagents and conditions required?

   [2]

   (c) The *E/Z* isomers of $C_2H_2Cl_2$ polymerise to form poly(1,2-dichloroethene).

   Draw a short section of poly(1,2-dichloroethene), showing **two** repeat units.

   [2]
   [Total: 9]

3 The flowchart below shows a reaction sequence starting from but-2-ene.

$$CH_3CH=CHCH_3 \xrightarrow{HBr} A \xrightarrow[\text{heat}]{OH^-(aq)} B \xrightarrow[\text{heat}]{Cr_2O_7^{2-}/H^+} C$$

   (a) Draw a short section of the addition polymer formed by but-2-ene showing two repeat units.

   [2]

   (b) Identify compounds **A–C** in the flowchart above.

   [3]

   (c) Name the types of reaction taking place in the flowchart above and state the role of the reagents.
   (i) $CH_3CH=CHCH_3 \longrightarrow A$

   (ii) $A \longrightarrow B$

   (iii) $B \longrightarrow C$

   [6]

   (d) What reagents and conditions would be needed to convert but-2-ene directly into compound **B**?

   [2]
   [Total: 13]

Practice examination questions *(continued)*

4 A branched alkene **A** was reduced with hydrogen to form a compound **B**. Compound **B** had the composition by mass of C, 82.8 %; H, 17.2 % (M_r = 58). Compound **A** was reacted with steam with an acid catalyst to produce a mixture of two structural isomers **C** and **D**.

(a) Calculate the empirical and molecular formula of compound **B**. [3]

(b) Show structural formulae for compounds **A–D**. [4]

(c) Write an equation for each reaction. [2]

[Total: 9]

5 (a) There are four structural isomers of molecular formula C_4H_9Br. The structural formulae of two of these isomers are:

- Compound **A**: $CH_3CH_2CH_2CH_2Br$
- Compound **B**: $(CH_3)_2CHCH_2Br$.

(i) Show the remaining two structural isomers.

(ii) Give the name of **B**.

[3]

(b) All four structural isomers of C_4H_9Br react with hot aqueous sodium hydroxide.

(i) What type of reaction is taking place?

(ii) Draw the structural formula of the organic product formed by the reaction of **B** with hot aqueous sodium hydroxide.

[2]

(c) Ethene, C_2H_4, reacts with bromine.

(i) State the name of the mechanism involved.

(ii) Show the mechanism for this reaction.

[4]

[Total: 9]

Practice examination answers

Chapter 1 Atoms, moles and reactions

1 (a)

particle	relative mass	relative charge	
proton	1	1+	✓
neutron	1	0	✓
electron	1/2000	1–	✓

[3]

(b) (i) neutron ✓; ^{13}C has 1 more neutron ✓

(ii) electron ✓; $^{16}O^{2-}$ has 2 more electrons ✓

(iii) proton ✓; $^{24}Mg^{2+}$ has 1 more proton ✓ [6]

[Total: 9]

2 (a) Difference: different numbers of neutrons. ✓

Similarities: same numbers of protons ✓ and electrons. ✓ [3]

(b) ^{11}B ✓ [1]

(c) (i) $1s^2 2s^2 2p^6 3s^2 3p^6$ ✓; (ii) p block ✓ [2]

(d) (i) $Si(g) \longrightarrow Si^+(g) + e^-$; ✓✓ (ii) The fifth electron is lost from a different shell ✓, closer to the nucleus ✓, less shielding ✓, experiencing a greater nuclear attraction ✓. [6]

[Total: 12]

3 (a) (i) 0.0170 ✓; (ii) 0.0850 mol dm⁻³ ✓✓;

(iii) 204 cm³ using 24 dm³; 211 cm³ using $pV=nRT$ (AQA) ✓✓✓ [6]

(b) 0.0741 mol dm⁻³ ✓✓✓ [3]

[Total: 9]

4 (a) (i) 0.0145 mol ✓; (ii) 0.0290 mol ✓; (iii) 23.2 cm³ ✓ [3]

(b) $M(Na_2CO_3)$ = 106 g mol⁻¹ ✓

mass Na_2CO_3 = 0.0145 x 106 = 1.537 g ✓ [2]

(c) Mg is oxidised from 0 to +2 ✓; H is reduced from +1 to 0 ✓ [2]

[Total: 7]

5 (a) (i) 0.0625 ✓; (ii) $M(NaN_3)$ = 65 g mol⁻¹ ✓;

moles NaN_3 = 0.0417 mol ✓; mass NaN_3 = 0.0417 x 65 = 2.71 g ✓ [4]

(b) (i) concentration = $0.0417 \times \dfrac{1000}{50.0}$ ✓ = 0.834 mol dm⁻³ ✓

(ii) Na is oxidised from 0 to +1 ✓; H is reduced from +1 to 0 ✓ [4]

[Total: 8]

Chapter 2 Bonding, structure and the Periodic Table

1 (a) A covalent bond is a shared ✓ pair ✓ of electrons. [2]

    (b) In a dative covalent bond, the same atom provides both electrons for the bonded pair ✓. The O of a water molecule uses a lone pair of electrons to form a dative covalent bond with H^+ to form an H_3O^+ ion ✓. [2]

    (c) H_2O: 104.5° ✓; H_3O^+: 107° ✓. In H_3O^+, there is repulsion between three bonded pairs of electrons and one lone pair of electrons ✓. The lone pair repels more than the bonded pairs, reducing the tetrahedral bond angle by 2.5° ✓. [4]

    (d) Hydrogen bonding ✓. This arises from dipole-dipole attraction ✓ between $H^{\delta+}$ of a water molecule and $O^{\delta-}$ on a different water molecule ✓. [3]

    [Total: 11]

2 Electron pairs repel one another ✓ and assume a position as **far away** from one another as possible ✓. Lone pairs repel more than bonded pairs ✓. Diagrams are shown below:

(a)   (b)   (c)

1 mark for each diagram
1 mark for each bond angle
1 mark for lone pair on PH_3

    ✓✓   ✓✓✓   ✓✓

tetrahedral ✓   pyramidal ✓   trigonal planar ✓   [Total: 13]

3 (a) (i) covalent ✓;   (ii) van der Waals' forces ✓. [2]

    (b) Weak van der Waals' (or induced dipole-dipole) forces break on vaporising ✓. [1]

    (c) Strong ✓ ionic bonds ✓ hold the lattice together. [2]

    (d) Ions vibrate faster ✓. [1]

    (e) On melting, the ionic bonds remain close together and are just loosened from their rigid lattice positions ✓. On vaporising, the ions must be separated, breaking the ionic attraction ✓. [2]

    [Total: 8]

4 (a) Water has hydrogen bonding ✓. This is dipole-dipole attraction ✓ between $H^{\delta+}$ of a water molecule and a lone pair ✓ on $O^{\delta-}$ on a different water molecule ✓. [4]

    (b) HCl has dipole-dipole interactions ✓ between $H^{\delta+}$ of a HCl molecule ✓ and $Cl^{\delta-}$ on a different HCl molecule ✓.   any 2 ⟶ [2]

    (c) Kr has van der Waals' forces only ✓. These are caused by oscillating dipoles ✓ which cause induced dipoles on another Kr atom ✓.
H-bonding > dipole-dipole > van der Waals' ✓. The intermolecular forces are broken on boiling ✓. The stronger the intermolecular forces, the higher the boiling point ✓.   any 4 ⟶ [4]

    [Total:10]

5 Across Period 3, the atomic radii decrease ✓ because there are more protons in the nucleus across the period ✓ increasing the attraction ✓ on the electrons in the same outer shell ✓.

Down Group 2, the atomic radii increase ✓ as more shells are being added ✓. The shielding effect increases as the number of inner shells increases ✓. The attraction experienced by the outer electrons from the nucleus decreases ✓.   [Total: 8]

6 (a) (i) $Ba(s) + 2H_2O(l) \longrightarrow Ba(OH)_2(aq) + H_2(g)$ ✓

       (ii) $2Mg(s) + O_2(g) \longrightarrow 2MgO$ ✓

       (iii) $Mg(s) + 2HCl(aq) \longrightarrow MgCl_2(aq) + H_2(g)$ ✓   [3]

    (b) (i)   On descending, outer electrons are further from the nucleus ✓ with greater shielding ✓. Therefore less attraction on outer electrons which are lost more easily ✓.

       (ii)   Reaction with H_2O/O_2/Cl_2/named acid ✓. [4]

    (c) (i)   Very high melting point/strong forces holding lattice together ✓.

       (ii)   CaO: cement ✓; $Mg(OH)_2$: indigestion remedy ✓ *many others possible.* [3]

    [Total: 10]

7 (a) On descending Group 7, the atomic radii increase ✓ resulting in less nuclear attraction at the edge of the atom ✓. There are more electron shells between the nucleus and the edge of the atom to shield the nuclear charge ✓. The nuclear attraction at the edge of the atom decreases down the group ✓. ∴ the oxidising power of the Group 7 elements decreases down the group ✓. [5]

(b) (i) $Cl_2 \longrightarrow NaCl$ (ox no: Cl 0 → –1) ✓ reduction ✓;

$Cl_2 \longrightarrow NaClO_3$ (ox no: Cl 0 → +5) ✓ oxidation ✓

(ii) 2.7 mol Cl_2 reacts ✓; 0.90 mol $NaClO_3$ forms ✓; 9.6 g ✓

(iii) $Cl_2 + 2NaOH \longrightarrow NaOCl + NaCl + H_2O$ ✓; bleach ✓ [9]

[Total: 14]

8 (a) (i) Electronegativity is a measure of the attraction of an atom in a molecule for the pair of electrons in a covalent bond ✓.

(ii) Electronegativity decreases from fluorine to iodine ✓. The atomic radii increase down the group ✓. The shielding effect also increases down the group as the number of electron shells increases ✓. The smaller the halogen atom, the greater the nuclear attraction experienced by the bonding electrons and the greater the electronegativity ✓. [5]

(b) The boiling points increase from fluorine to iodine ✓. The number of electrons increases ✓, increasing the van der Waals' forces between molecules ✓. [3]

(c) With aqueous potassium chloride, there is no change ✓. With aqueous potassium iodide, the solution turns dark brown ✓ as iodine is displaced ✓.

$Br_2(aq) + 2I^-(aq) \longrightarrow 2Br^-(aq) + I_2(aq)$ ✓ [4]

(d) Bromine is a weaker oxidising agent than chlorine because it does not oxidise chloride ions to chlorine ✓. Bromine is a stronger oxidising agent than iodine because it oxidises iodide ions to iodine ✓. [2]

[Total: 14]

Chapter 3 Energetics, rates and equilibrium

1 (a) $2Na(s) + C(s) + 1\frac{1}{2}O_2(g) \longrightarrow Na_2CO_3(s)$ ✓✓ [1, equation; 1, state symbols] [2]

(b) 100 kPa and 298 K ✓ [1]

(c) +90 kJ mol^{-1} ✓✓✓ [3]

[Total: 6]

2 (a) (i) The enthalpy change that takes place when one mole of a substance ✓ reacts completely with oxygen ✓ under standard conditions, all reactants and products being in their standard states ✓.

(ii) 298 K ✓ [4]

(b) (i) $CH_3CH_2CH_2OH(l) + 4\frac{1}{2}O_2(g) \longrightarrow 3CO_2(g) + 4H_2O(l)$ ✓

(ii) –2023 kJ mol^{-1} ✓✓✓ [4]

[Total: 8]

3

(a) The enthalpy change that takes place when one mole of a compound ✓ in its standard states is formed from its constituent elements ✓ in their standard states under standard conditions ✓. [3]

(b) $C(s) + 2H_2(g) \longrightarrow CH_4(g)$ ✓✓ [1, equation; 1, state symbols] [2]

(c) –76 kJ mol^{-1} ✓✓✓ [3]

(d) C–H: +412 kJ mol^{-1} ✓; C–C: +348 kJ mol^{-1} ✓✓ [3]

[Total: 11]

4 (a) (i) ✓✓✓; (ii) ✓✓

[5]

(b) (i) ✓✓✓; (ii) ✓

(iii) A catalyst provides an alternative reaction route ✓ with a lower activation energy ✓. More molecules exceed the lower activation energy so more are able to react. ✓ [7]

[Total: 12]

5 (a) (i) A homogeneous equilibrium has all equilibrium species in the same phase ✓. All species in this equilibrium are gaseous ✓.

(ii) The forward reaction proceeds at the same rate as the reverse reaction ✓. The concentrations of reactants and products are constant ✓. Equilibrium can only be achieved in a closed system ✓. [5]

(b) (i) Equilibrium position is unaffected ✓. There are equal numbers of gaseous moles on either side of the equilibrium ✓.

(ii) Rate increases ✓. Increasing the pressure also increases the concentration ✓. [4]

(c) Endothermic ✓. With increased temperature, an equilibrium shifts in the endothermic direction ✓. [2]

[Total: 11]

Chapter 4 Organic chemistry, analysis and the environment

1 (a) (i) Same **molecular** formula with a different arrangement/structural formula ✓.

(ii) $ClCH_2CH_2CH_2Cl$, $CH_3CHClCH_2Cl$, $CH_3CH_2CHCl_2$, $CH_3CCl_2CH_3$ ✓✓✓ [5]

(b) (i) UV / sunlight ✓; (ii) C_3H_7Cl ✓

(iii) $CH_3CH_2CH_2Cl$ ✓ 1-chloropropane ✓
$CH_3CHClCH_3$ ✓ 2-chloropropane ✓

[6]
[Total: 11]

Organic chemistry, analysis and the environment *(continued)*

2 (a) (i) There must be a C=C double bond that prevents rotation. ✓
There must be different groups attached to each carbon atom in the C=C double bond. ✓
(ii) *E*: groups are on opposite sides of double bond.
Z: groups are on same side of double bond. ✓
(iii) correct structures ✓✓

Z   E   [5]

(b) H_2 ✓ and a named catalyst/Ni/Pt/Pd ✓.   [2]

(c)

2 repeat units ✓ with side links ✓   [2]
[Total: 9]

3 (a)

no double bond ✓, correct skeleton ✓   [2]
(b) **A** = $CH_3CH_2CHBrCH_3$ ✓; **B** = $CH_3CH_2CHOHCH_3$ ✓; **C** = $CH_3CH_2COCH_3$ ✓   [3]
(c) (i) addition ✓, electrophile ✓;
(ii) substitution ✓, nucleophile ✓;
(iii) oxidation ✓, oxidising agent ✓.   [6]
(d) steam ✓ and H_3PO_4 ✓   [2]
[Total: 13]

4 (a) Empirical formula of **B** = C_2H_5 ✓✓; Molecular formula of **B** = C_4H_{10} ✓   [3]
(b) **A** = $(CH_3)_2C=CH_2$ ✓; **B** = $(CH_3)_2CHCH_3$ ✓;
C and **D** = $(CH_3)_2CHCH_2OH$ ✓ and $(CH_3)_2COHCH_3$ ✓;   [4]
(c) $C_4H_8 + H_2 \longrightarrow C_4H_{10}$ ✓
$C_4H_8 + H_2O \longrightarrow C_4H_9OH$ ✓   [2]
[Total: 9]

5 (a) (i) $CH_3CH_2CHBrCH_3$ ✓; $(CH_3)_3CBr$ ✓.
(ii) 1-bromo-2-methylpropane ✓   [3]
(b) (i) nucleophilic substitution ✓; (ii) $(CH_3)_2CHCH_2OH$ ✓   [2]
(c) (i) electrophilic addition ✓
(ii) arrows stage 1 ✓; arrow stage 2 ✓; carbocation ✓

[4]
[Total: 9]

The Periodic Table

Index